# The Reform of the CAP and Rural Development in Southern Europe

T0186674

*Edited by*

CHARALAMBOS KASIMIS
*Department of Economics*
*University of Patras*
*Patras, Greece*

GEORGE STATHAKIS
*Department of Economics*
*University of Crete*
*Rethymno, Greece*

ASHGATE

Published by
Ashgate Publishing Limited
Gower House
Croft Road
Aldershot
Hants GU11 3HR
England

Ashgate Publishing Company
Suite 420
101 Cherry Street
Burlington, VT 05401-4405
USA

Ashgate website: http://www.ashgate.com

**British Library Cataloguing in Publication Data**
The reform of the CAP and rural development in Southern Europe.
  - (Perspectives on rural policy and planning)
  1.Rural development - Europe, Southern 2.Agriculture and
  state - Europe, Southern 3.Europe, Southern - Economic
  policy
  I.Kasimis, Charalambos II.Stathakis, George
  307.1'412'094

**Library of Congress Cataloging-in-Publication Data**
The reform of the CAP and rural development in Southern Europe / edited by Charalambos Kasimis and George Stathakis.
      p. cm. -- (Perspectives on rural policy and planning)
  Includes bibliographical references.
  ISBN 0-7546-3126-5 (alk. paper)
  1. Rural development--Europe, Southern. 2. Agriculture and state--Europe, Southern.
  3. Europe, Southern--Rural conditions. I. Kasimis, Charalambos. II. Stathakis, George.
  III. Series.

  HN380.Z9 C6667 2003
  307.1'412'094--dc21

                                                          2002028130

ISBN 0 7546 3126 5

Printed and bound in Great Britain by MPG Books Ltd, Bodmin, Cornwall

# THE REFORM OF THE CAP AND RURAL DEVELOPMENT IN SOUTHERN EUROPE

# Contents

# List of Tables

# List of Contributors

**Nikos Beopoulos**, Department of Agricultural Economics and Development, Agricultural University of Athens, Greece.

**Jean-Paul Billaud**, CNRS-LADYSS, France.

**Karl Bruckmeier**, Department of Human Ecology, University of Goteborg, Sweden.

**Michael Demoussis**, Department of Economics, University of Patras, Greece.

**Francesco Di Iacovo**, Department of Agricultural Economics, University of Pisa, Italy.

**Fernando Garrido-Fernández**, IESA, CSIC, Cordoba, Spain.

**Chris Gerry**, Department of Economics and Sociology, University of Tras-os-Montes e Alto Douro(UTAD), Portugal.

**Charalambos Kasimis**, Department of Economics, University of Patras, Greece.

**Terry Marsden**, Department of City and Regional Planning, Cardiff University, United Kingdom.

**Eduardo Moyano-Estrada**, IESA, CSIC, Cordoba, Spain.

**Teresa Patricio**, Ministry of Science and Technology, Portugal.

**José Portela**, Department of Economics and Sociology, University of Tras-os-Montes e Alto Douro (UTAD), Portugal.

**George Stathakis**, Department of Economics, University of Crete, Greece.

**Kostas Vergopoulos**, Department of International and European Studies, Panteion University, Greece and University of Paris VII, France.

# Preface

The idea for the publication of this volume originates from a Conference entitled 'New Policies for the Development of Countryside in Southern Europe'. The Conference was organized in Athens in November 1998 by the 'Nicos Poulantzas' Foundation and the Institute of Urban and Rural Sociology of the National Centre for Social Research of Greece (EKKE).

This book discusses the various aspects connected with the Reform of the CAP and the implementation of the new rural development policies. European countryside has been undergoing a significant transformation connected with new forms of production and consumption, and increasing concerns related to environmental sustainability. These concerns, encapsulated in Agenda 2000, constitute a policy shift that poses a number of social, economic and political challenges with respect to the reality of Southern European countryside.

The CAP and the rest of implemented regional-structural policies had uneven results in the Southern European member states, which still maintain a large agricultural sector. As a result, Southern European countries now face the challenge of adjusting to a new political conjuncture. This conjuncture, however, is realized within a global context significantly determined by the liberalization of world markets and the enlargement of the European Union. The restricting factors and the prospects of these adjustments for the countries of the European South are some of the issues discussed in this book.

We gratefully acknowledge the support of the 'Nicos Poulantzas' Foundation in the preparation of this volume. Special thanks go to Maria Fokou for her technical work on the formatting of this book and to Demetra Fannie Kasimis for her editorial help. Additionally we are grateful to colleagues D. Damianos and D. Psaltopoulos who read an earlier version of the Introduction and made valuable comments. Finally we would like to thank all the contributors to this publication who patiently worked with us over the past year.

C. Kasimis
G. Stathakis

Chapter 1

# Introduction and Overview of the Volume

Charalambos Kasimis and George Stathakis

## New Challenges for Rural Southern Europe

For several years now, approaches to rural development have distanced themselves from the more traditional approach, which tended to identify the countryside with agriculture, and have focused more on an analysis of the prevailing sector – specific (economic, social, and cultural) relations (Marsden, 1995). Today the use of the term *countryside* reflects this dual historical shift. The first shift is due to the decreasing importance of agriculture – both in terms of employment and production – in European economies (Gardner, 1996). With this development and the relative stabilization of the rural population, an increasing part of rural inhabitants is being drawn into the non-agricultural sectors such as tourism, construction, manufacturing, and the conventional and innovative services, thereby granting a more diversified and contemporary role to the countryside (Boyle and Halfacree, 1998). The second shift is toward the environment. The environmental shift has questioned many of the very fundamental premises governing the relation between social practices and nature. It has challenged the conventional attitudes that focus on the economic performance of the agricultural sector and perceive farming as a typical industrial activity (Drummond and Marsden, 1999).

An analysis of the changes in the countryside is bound to be incomplete unless it incorporates the wider configurations that have reshaped developed economies over the last two decades. The broader and more intensive use of market mechanisms, the liberalization of national economies and the accelerated circulation of international capital and commodities have brought about profound changes in terms of production and consumption patterns. At the same time, the adoption by most countries of a common monetary and fiscal policies base while concomitantly disengaging from sectoral policies and planning has drastically altered the modes of regulation and management of the economy. These trends are usually related to a wider technological revolution that seems inevitably to force economies to adopt specific production and employment patterns. All the while, enterprises seem to be pushed toward either larger business conglomerations or smaller, but flexible, entrepreneurial units.

Agriculture seems no exception to these wider trends. The Uruguay Round Agreement on Agriculture and the reform of the Common Agricultural Policy

(CAP) show movement toward more market-oriented arrangements. As a result, the further reduction of tariffs and subsidies seems inevitable. Such developments were expected to favour the existing pattern of growth of multinational corporations in the agro-food network (Friedland et al., 1991) and the rapid and controversial expansion of biotechnology (Goodman, 1999). These processes, however, occur even while the locally based food production and consumption networks that favour diversification, quality production and the regeneration of small productive units in a plethora of specialized regional markets continue to affirm themselves.

Furthermore, Agenda 2000 and the recent and ongoing debate concerning the CAP reforms are connected to a large extent with these new realities (Brower and Lowe, 2000). Such policy changes can only be partially attributed to the pressures created by globalization processes. The internal dynamics of change in rural areas are of equal importance. The transformation of the countryside into consumption space with a diversified composition of production and income permits the strong belief that agriculture can become part of the broader process of setting the market as the main mechanism for allocating resources and determining production outcomes.

In addition, the gradual erosion of regulatory and planning policies experienced in most European states during the last two decades, have brought on profound shifts. The issue of 'governance' of the rural space may seem to constitute an exemplary shift toward more diversified, local and democratic forms of planning, but, at the same time, it raises the question of the magnitude of resources that are to be committed to objectives related to regionally sustainable development.

Within the context of these changes, Southern European rural areas seem to encounter the greatest challenges. Lagging behind in comparison with Northern European's agricultural size, technology and organization, Southern European agriculture is forced to adapt in this competitive arena, while undergoing a shift toward sustainable development. In this competitive arena increases in productivity and improvements in quality are pursued, and additional sources of income possibilities are investigated. Systems of production in Southern European countrysides remain very diversified, but in many regions the local socio-economic fabric continues to depend extensively on agriculture; whereby potential changes are a cause of great concern (Damianos et al., 1998). As direct subsidies are being reduced and tariff protection decreases, part of the agricultural sector is threatened not only by cheap imports but also by innovative and technology-powered competitive agriculture.

At the same time, the advantages derived from these shifts usually depend on complex systems of institutional and governing bodies. The demand for a 'new governance' of the rural space imposes challenges for the less developed regions that can hardly be met. With a long tradition of centralized state-management, tight control over local administration, and a looser social weave (compared to the Northern European countries), the South must, in addition, build institutions, incorporate planning practices and introduce innovation-oriented processes.

As in every transitional period abrupt changes are viewed as either a threat or a challenge. In such cases, social scientists impulsively look for breaks and turning points. Changes, however, tend to obscure continuities and stability across time and undermine processes that tend to shape and reshape diachronic constant patterns. Historians often seek refuge in the Braudelian, *longue durée*, methodological term (Braudel, 1969). The term implies an attempt to study continuities and regularities, and place changes, transformations, and realignments within its context. Accordingly, no transformation is exogenous to a system – whether it be productive, social, or economic. The elements of every transformation are very much present within all existing systems, long before they manifest themselves in the shape of some profound shift.

This approach might be useful in analysing current developments in the countryside. Two sketches are outlined here; the first focuses on the patterns of stability and the second on the patterns of change in the European countryside. The discussion then moves to the policy debate, i.e. the evaluation of current policy reforms and the search for alternative policies for the countryside.

**The Patterns of Stability**

Several stable trends can be detected in post World War II rural Europe. The consistent decline in European countries of the share by the agricultural sector in terms of both employment and the Gross National Product (GNP) despite improved production techniques and increased output results has been accompanied by an increased performance in both entrepreneurial (capital intensive) and traditional (labour intensive) agriculture, even though there has been a significant decrease in the number of people employed in European agriculture. In addition, despite the delayed phenomenon of market saturation, or surplus production in many sectors, production growth over the last fifty years seems to have gone hand in hand with the slow increase rate of demand (Fennell, 1997).

A second element of stability is that of agricultural policy. National policies, and (to a certain point) the CAP, have shaped a policy framework which has remained, despite its gradual changes, exceptionally stable over time (Fennell, 1997; Fearne, 1997). Such a framework is made up of a combination of (more or less) common policy measures: price support, intervention and tariff protection, export subsidies, public financing of infrastructure and state-subsidization of technological innovations.

One hypothesis is that these two patterns of stability are related. The policies that have prevailed for so long have protected the agricultural sector from sudden swings and extensive restructuring phenomena which, at times, (mainly during recession periods) have marked the rest of the economy.

According to the same perspective, two other areas of stability can be underscored. First, the actual division of agricultural production between the northern and the southern regions of Europe, which has determined intra-European trade, has remained relatively unchanged. In this respect, northern regions have retained their specialization in livestock production and certain extensive cultivation groupings of products, while southern regions continue to specialize in Mediterranean products. Second, the dominant types of production units in the

North and the South of Europe (i.e. more entrepreneurial in the former and more reproductive in the latter) have remained relatively unchanged over time. In both cases, both the forms of land ownership and production structures have proved stable. These forms, without downplaying the changes to be discussed below, do take on importance. Technological shifts, as impressive as they might have been, and increases in rural income (with the resulting changes in the consumption habits of the rural population) failed to alter the framework of inherited social relations in any drastic way. Furthermore, they have not displaced long-established forms of institutional and economic management of the agricultural economies.

## The Patterns of Change

Over the same time period examined above, after a significant decrease in the population of the countryside in the 1950s and 1960s, it levelled off and, in some cases, increased. The countryside became a place of new production sites, residential expansion, and new consumption activities (climate, topography and natural amenities contributing to rural increases in population and employment. Thus, broad semi-urban zones were created and the very boundaries of the countryside became questionable.

Regionally diversified, these processes were usually and relatively more intense in coastal, rather than inland areas, in plains, rather than mountainous regions. In some (mainly Northern European) countries, these changes were noted as early as the 1960s, while in others (mainly Southern European) countries, such transformations were extensively realized only in the 1980s and 1990s.

This new non-homogeneity has fundamentally disturbed previously accepted perceptions of the countryside as a more or less homogeneous agrarian society. This phenomenon was well recorded in various local empirical studies of a local nature that flourished during the 1980s and 1990s and indicated, with certainty, the heterogeneity and multiplicity of modern economic formations in the countryside (Bradley and Lowe, 1984; Fuller, 1990; Kasimis and Papadopoulos, 1994; Miller, 1996).

In the new reality, rural social class structure has become more difficult to analyze and interpret. Apart from the usual distinctions among farming households, new social groups have emerged that diverse according to the type of economic activities developed in their region. Forms of employment have varied significantly; pluriactivity has become more widespread (Barlett, 1986; Gasson, 1988; Kayser, 1990), new middle classes have been formed, and non-family labour, mainly immigrant, has increased extensively (Rees et al., 1996; King et al., 1997; Hoggart and Mendoza, 1999).

At the same time, new tensions within local communities and regions rising from new controversial issues of land use, alternative uses of water and other natural resources, location of economic activities and the management of settled environment have entered the scene. This is so particularly in Southern European countries, where these tensions have served as catalysts for change, due to the enormous pressure that the explosion of tourism and the sustained incursions that summer, legally or illegally constructed, residences have exerted on resources (Papadopoulos, 1999). In turn, such tensions are indicators of the need for a radical re-evaluation of the countryside, and more importantly, a sign of new policies for rural development.

## The Limitations of the CAP

It seems unnecessary to evaluate the scopes, objectives and impact of the CAP here. Of greater interest is the need to examine the environmental implications of the productivist model supported by the CAP. The critical arguments are more or less familiar (Whitby, 1996; de Haan et al., 1997). Intensive agriculture is considered self-destructive over the long run and the suggestion is that agricultural production should not be directed by shortsighted, purely economic, reasoning. Such production has to do directly with the relationships underlying the economy and nature and inevitably involves environmental, social, and, in particular, ethical issues. The environmental perspective demands a reversal of priorities with calls for the revival of local markets, the diversification of local production, and the disengaging of agriculture from the output goals of steady economic growth.

From the very beginning, such a perspective brought back the issue of resource management and the aims and methods of agricultural planning, and suggested the adoption of environmental regulatory mechanisms. It urged, in contrast to central bureaucratic management, the building of new local institutions (and forms of local action) and emphasized the importance of local democracy and the decentralization of the political decision making process. Thereby, wider theoretical and political traditions such as communitarianism, direct democracy, self-management, ecology, social economy, and endogenous and sustainable development converged (Friedmann, 1987; Goodwin, 1998; Ray, 2000; Roseland, 2000).

This heterogeneous set of ideas had a significant outcome, and manifested itself in its incorporation in the new rural development policy embodied in Agenda 2000 in which such rural development has itself integrated into the CAP as a 'Second-Pillar'. Rhetorically speaking, most institutions fully identified themselves with such concepts. Sustainable development now found itself promoted as a necessary parameter in all kinds of studies, projects and policies. Practically speaking, however, results produced to date have tended to be meagre, having involved actions of only limited significance. A true application of policies derived from sustainable development would be, it seems, in direct conflict with the fundamental tenets of the CAP.

Yet, this gap between rhetoric and actuality became even more apparent in discussions related to the liberalization of international markets. At a time of discreet inter-nationalization of the economy and revival of liberal thinking, the most protected sector that of agriculture, could no longer be excluded from such discussions. The Uruguay Round Agreement, the series of discussions of the World Trade Organization (WTO) (Tangermann, 1999), the current changes of the CAP and more importantly the continuing discussions concerning its fundamental transformation, have already set the course of change. In opposition to the regime of regulations and interventions (although the degree and intensity of changes remains under discussion), a relative increase of the role of the markets seems inevitable.

The idea of developing entrepreneurial agriculture relating it to the market has found support in the CAP since the early 1980s. In recent years, entrepreneurial agriculture in countries of the Southern Europe has gained significant ground, unlike most Northern European countries where the diversification of consumption patterns have encouraged initiatives in areas of production outside the CAP regimes. In these sectors entrepreneurial agriculture showed rapid development, indicating that market mechanisms may factor well in adjusting production initiatives to new consumer needs.

Taking into consideration the particularities of the agricultural sector, the idea of putting into practice the rules of the free market seemed strange at first. The adoption of such rules would legitimize their outcome. According to these rules, the absence of competitiveness would lead to bankruptcy, and positive competitiveness would result in expansion and dominance. Consequently, extensive restructuring of production would be inevitable, and many types of non-competitive production units would disappear. This does not necessarily imply a homogeneous production structure; but it does mean that marginal production would be seriously tested, and extreme heterogeneity in production would be drastically reduced. Any attempt to reduce the prices of agricultural products through increased competition would bring about not only more specialization and domination of intensive forms of farming, but also increased pressure on the environment (Redclift et al., 1999). In addition, it would put the particularities of deeply rooted social relations in the agrarian countryside at risk.

On an international scale this undertaking carries new threats. A fully organized agriculture, like that of the USA, stands clearly superior to that of European agriculture, and even more so, to agriculture in most developing countries. Not only does this threaten the economic stability of countries in which agricultural exports have a predominant position in their external trade, but it also reveals the constraints posed to the international economy, which so far has been unable to secure a balanced development of its constituent parts.

Furthermore, it would be absurd in the liberalizing of international markets to bring about an end to all forms of national support to agriculture. On the national level, support systems for agricultural production, whether direct or indirect, will more than likely continue (Robinson, 1994). The danger arising from the liberalizing of international markets is that pressures from international competition would translate as pressures on national fiscal policies. Likewise, every loss in international competitiveness would have a domestic financial cost.

Transferred into the European arena this idea would imply an extreme revision of the CAP, i.e. the renationalization of agricultural policy. Though the consequences of such a development might turn out to be gentler in comparison with those at the international level, they would, nonetheless, bring into play the abovementioned contradictions. In addition marginal Mediterranean agriculture, despite its record of resistance to marginalization (Bazin and Roux, 1995) would face enormous pressures in those sensitive, poor-performance regions even within the current support system.

The pressure of these developments on Europe is quite obvious. The European Union (EU) is expected to face the cost of its expansion towards Central and Eastern Europe, while Southern European countries will continue to put pressure on the EU for various forms of support for their economies. There are demands stemming from the introduction of the common currency and the urgent need to tackle unemployment. All these create insurmountable pressures on the European budget. As long as the budget remains at its current level (just 1.28 per cent of the European GNP), fiscal pressures cannot be accommodated.

The rural policy debate should take primary importance in talks concerning the formulation of strategies for resolving some very basic and contradictory issues characterizing the present day countryside. The new policies must balance the insistent demands toward a more production oriented rather than an environmentally sensitive agriculture and take into account the pressures put by demands on countryside as both a production and consumption space. They need to integrate the several dimensions of multi-sectoral production and employment structures within a framework that addresses economic and social objectives (and thereby avoids the marginalization of certain social groups and neglect of less developed regions). They need to address and integrate the different needs of the Northern Europe with the needs of the Southern Europe and satisfy the expectations of prospective members involved in this process of enlargement. Finally they have to allow more market-oriented policies to coexist with more extensive regulatory systems addressing their social and environmental concerns.

## Overview of the Volume

This book is about the prospects for the development of Southern European countryside during a transitional period of a major policy paradigm shift. The move from a stable CAP framework to a new, but incomplete, rural development environment has opened up a new research and policy agenda.

The first part of this book presents four papers that focus on developments in the European countryside, the CAP reform and the policies of the 'Second Pillar'. These topics provide the general context within which other issues may be discussed. In the second part, five papers are presented that explore the cases of Portugal, Spain, France, Italy and Greece. Each paper places different emphasis on some of the issues of the research agenda mentioned above.

The volume begins with an introductory paper by Marsden. This paper sets the overall framework necessary for comprehending both the actual developments in European countryside and the shifts in policy. It attempts to capture, in theoretical terms, the contradictory nature of the policy changes as initiated by the different perceptions of countryside. Marsden's basic argument is that the governance of rural space in the European context is unsustainable. In terms of growing concerns on health, environment, welfare, and employment issues, the current system of agricultural and rural regulation in Europe has become obsolete. It faces a long-term crisis of legitimacy among both the rural population and the urban consumer classes. The CAP remains a highly asymmetrical support system. Despite numerous policy initiatives towards a sustainable rural development approach to the European countryside, funds continue to be directed predominantly toward policies that follow the agricultural productionist model.

Marsden identifies three social, political, scientific, and economic dynamics (agro-industrial system, post-productionist model, sustainable development) currently combining and competing to shape European rural space. They are all variably infused into the current policy debate concerning rural development.

The agro-industrial system tends to see the reform of the CAP as a preface for further concentration in the agro-food sector and is associated with attempts to create a global market in which agri-business and intensive producers can compete.

The post-productionist model, identified with the new urban strata that move deeper into the rural fabric, has turned much of the receiving countryside into 'consumption spaces'. While the agro-industrial model tends to marginalize rural nature through the production process, the post-productionist model does the same through a highly materialist conception of the consumption process.

The model of sustainable development is marked by a different set of organizing principles. Capital is reconfigured in a more symmetrical relationship with labour and land, and emphasis is placed upon a more multi-functional, land-based, production organization. It is diversity rather than convergence that is important. Such an 'alternative system' requires new forms of quality regulation, making it increasingly difficult for national governments to control and regulate.

Ultimately, Marsden turns to a discussion of how regulatory and governance frameworks may be developed so that sustainable rural development gains momentum. He indicates the need of building alliances from within – at least at the regional and local levels. The governance of rural space, however, is likely to become even more involved in the redefinition and re-evaluation of nature. This will take it far beyond the simple exploitative modus operandi of the agro-industrial or post-productionist models of rural development.

Beopoulos presents the features and variety of Mediterranean farming systems and indicates the differences between Northern and Southern European agriculture. He analyzes the way Mediterranean products are dealt with within the CAP and then traces the possible consequences for Mediterranean products in light of the Agenda 2000 proposals.

He identifies three distinct periods of the CAP. First, the CAP's consolidation period of 1970-1980; following this, the period of enlargement towards the Mediterranean, when the issue of convergence and cohesion was set, leading to the introduction of the structural programmes; and finally, the period after the 1992 reform, when CAP moved towards new horizontal orientations (eco-conditionality, employment, rural development) that signified the consolidation of the idea of multifunctionality of agriculture.

In the context of the multifunctionality of agriculture, certain possibilities open up for products associated with the Mediterranean region. The Mediterranean region of the EU, rich in extensive production systems, product variety, and a reserve of tastes and 'know-how', could utilize spaces, production models, and the gastronomic dimension of products, which might otherwise, in the name of progress, be called archaic. Issues pertaining to rural space and the role agriculture could play in this context are many, but as of yet, the new CAP orientations have not formulated clear and satisfactory solutions for most of them.

Beopoulos welcomes the prospects created by Agenda 2000 for an agricultural policy that concedes a significant role to national and local authorities. These plans could be advantageous for the South because of the possibilities they offer for the adjustment of policy to the needs of a specific area, and the room for manoeuvring the rigid regulations applied to the whole of Europe. Yet, at the same time, the risk of a reduction in EU funds intended for the development of less favoured areas and the typical ineffectiveness of public administration in many areas of the South might ultimately undermine this process.

Bruckmeier and Patricio deal with the agri-environmental policy that has become part of the CAP since 1992, and was designed to support four types of environmentally sound agriculture: traditional forms of agriculture; extensive agriculture; integrated farming; and organic farming. So far, the authors claim that agri-environmental policy is an experiment more of political aspiration than of widespread social practice.

They differentiate the models of 'sustainable transition' as two types of sustainable development approaches. The first approach describes the continuity of 'policy as usual' in a state-controlled, interventionist model. It favours a 'top-down approach' and the construction of standardized rural development policies. The second favours micro-policies, allows for a better consideration of regional contexts and social actors, and often merges development strategies with simplified forms of 'bottom-up approaches' and participatory development. The critical question for both approaches is that of the coherence between processes at different levels of policy design, called 'the fit between ecosystems and social systems', and according to the authors, such coherence may vary considerably from one policy to another.

The authors then evaluate the impact from the introduction and implementation of the agri-environmental measures of the CAP reform in four countries: France, Germany, Portugal, and Spain. The comparison of the evidence from these four cases indicates different ways of response. The four cases can be seen as illustrative of the dominance of socio-cultural effects (France), economic effects (Germany), socio-economic policies of stabilizing traditional agriculture (Portugal), and institutionalized contradictions between ongoing mainstream modernization and ecological adaptation (Spain).

Bruckmeier and Patricio remain highly critical of such policies. Firstly, agri-environmental measures do not represent an optimal 'policy mix' but more a box of multi-purpose tools with too many diverse and overly ambitious objectives. Secondly, the agri-environmental measures have little if any impact on intensive and highly specialized agriculture. Thirdly, the strengthening of traditional farming systems has not been due to the effects of agri-environmental measures. Rather it is a synergetic effect of other factors such as support from local actors and groups, ecological movements and non-governmental organizations.

Finally, they indicate that there are unsolved problems that must be overcome before agri-environmental policy may become an effective tool for sustainable development. These include scientific questions, methodological problems referring to the socio-political, economic and ecological parameters of agricultural production systems and a whole variety of complex policy issues. They indicate the example of regional experiences in coastal areas from which much can be learned. Agri-environmental policy is in danger of ending up supporting an 'inner periphery' of rural areas within the countries of the EU. If this happens, rural development would become an example of the negative variant where social and economic marginalization can turn into 'ecological marginalization'.

Vergopoulos approaches the new developments in the rural sector through the perspective of European economic cohesion. He recalls the theories of economic integration according to which stability of the whole presupposes the adjustment of its components.

For Vergopoulos, Europe has, since the 1960s, acknowledged the issues of stability, harmonization, and integration of European markets. In the end, the CAP has turned out to be nothing more than a system of stabilizing price intervention aimed at balancing regional disparities contributing to the cohesion of the large internal European market and the structural reform necessary for long-term convergence.

Toward the end of the 1980s, liberal deregulatory policy catalysed European incorporation into the new liberal order of international trade through the gradual abolishment of any price regulatory systems. The CAP liberal reforms of 1992 introduced the principle of direct agricultural income subsidies to counterbalance the decrease in prices resulting from the disorganization of European agricultural markets. The final account, even if it was not as destructive as expected, given that incomes were kept at similar levels through subsidies, was unsettling. The main cause of this was a halt in the consolidation process of European cohesion and integration.

Vergopoulos goes on to criticize at length the idea of the renationalization of agricultural policy that is contained in Agenda 2000. With this, he argues, European agriculture is straying away from European integration and is being reinstated to its pre-integration national level. Today, in the context of globalization and deregulation, European agriculture is being 'de-europeanized' and reincorporated into a national framework.

The author ends up analyzing the contradictions arising from the parallel processes of integration at the international and European levels. Europe, by failing to maintain stability and cohesion of internal markets and to challenge unemployment, is selling out two decisive issues that legitimize Europe in public opinion. Convergence models formed to suit arguments in support of international market integration, in substance destroy the prospect of European integration. The immediate consequence is that in the resulting gap the old national complementarities and the antagonisms amongst states – not only on the economic and political issues but also on the social and cultural, too – will be intensified. Given this prospect, true dangers lie ahead not because of historical innovation or the uncertain course followed thereafter, but for exactly the opposite reason: the lack of any innovation and the repetition of realities and situations already known from the past.

Portela and Gerry set the stage for the discussion of Mediterranean rural development first by stressing the considerable differences between Southern and Northern European agriculture. The authors then discuss how the CAP has exacerbated the specific differences. They focus on the broader economic, political and social context that conditions the functioning and performance of agriculture. They point out four such areas: the experience of industrialization, information and knowledge networks, the physical and telecommunications infrastructure, and the welfare state. The authors argue that these factors have made Mediterranean economies particularly vulnerable to the demands of competitiveness, adjustment and economic restructuring, whether imposed by economic crises, globalization, and/or by the demands of European convergence.

A part of the paper is devoted to sketching out the basic contours of a re-engineered CAP. In order to achieve a more balanced development of European rural areas, two intermediate targets – socio-economic cohesion and a sustainable environment – would have to be met. This could best be done by operationalizing four guiding principles: putting people before commodities, and social cohesion before markets; giving due recognition to multi-functional farming and occupational pluralism; promoting food quality through locality; and imposing extensive eco-conditionality.

Portela and Gerry stress that the architecture of a re-engineered CAP would require three levels of institutional and policy initiatives. From a structural viewpoint, such a policy could be envisioned as consisting of a three-tiered process, composed of what they call unionization, macro-regionalization and nationalization, each of which would operate in line with the principles presented above. In effect, Portela and Gerry argue that agriculture should be brought back to the centre stage of regional development. They want to avoid the risk of seeing agriculture defined once and for all as the exclusive preserve of multinationals and the large and wealthy farmers, with the notion of rural development relegated to a series of residual policies providing a welfare safety net against socio-economic exclusion for pluriactive smallholders.

In the next paper Moyano and Garrido argue that the process of transformation characterizing western economies has far-reaching consequences for cultural and political value systems. Many of the principles on which government policies were based in the past are being revised. In response to such concerns, successive reforms of the CAP have been carried out since 1992.

The authors focus on the analysis of the way in which these changes are perceived within Spanish rural society where change is not homogeneous, but differentiated. New social groups have emerged in parallel to the diversification of the old social groups and as a result rural society has become very complex and diversified.

Next they discuss, in detail, the responses of the various social groups and the organized interests. The traditional corporatist scenario, based on the principle of common interest among farmers, is being replaced by one of plurality. This is reflected in the diversity of discourses, strategies, and options for the various organizations of the agricultural sector. Regarding agricultural co-operativism there is a fairly homogeneous perception of change, strengthening the enterprise-oriented criteria in the management of co-operatives. Such economic parameters predominate over the old principle of unity based on a sense of belonging to a social movement. On the contrary, many of the typical traditions of co-operativism can be found among the new groups of farmers that have emerged from the establishment of alternative systems of agricultural production and usually function through socio-economic networks.

In the case of farmers unions, two types of responses can be distinguished. First, the 'enterprise-oriented response', espoused by those organizations that primarily represent the interests of medium and large-scale farms. The second type of response could be described as 'neo-peasant', because it emphasizes the renewed values of rural society and provides the key role to family farming. This response is voiced by organizations representing the interests of small farmers. In relation to agricultural workers' unions, there is an 'adaptive and reformist response' and a 'rupturist and radical response'. The 'reformist response' has shifted emphasis from the question of land ownership to that of wages and welfare. The 'radical response' continues to question the very basis of the current structure of land ownership. Moyano and Garrido argue that the present context of change offers opportunities to all these groups to revitalize rural society, allowing them, by their participation in local-policy making, to influence decisions taken at the local level.

In such a context, old agricultural policies directed to market regulation and structural modernization are being reformulated according to the paradigm of sustainability and ecological modernization, and new policies are emerging to regulate the many functions of the countryside. Old and new policies and protagonists coexist in this period of transition and give Spanish rural society an unprecedented dynamism, offering new opportunities to different social groups.

Billaud focuses on negotiated development investigating whether that may lay the foundation for new forms of governance of rural societies. Negotiation procedure became the main process by which social actors were asked to take concerted action. Negotiated development involves different parties both from the public and the private sector faced with increasingly complex problems in the decision-making process. In every case, the determination to make consultation an integral part of the decision-making process is manifest. Thus it became necessary for all parties involved to mobilize expert opinion to provide a legitimate basis for public policy. This recourse in turn raises the question of the relationship between a technical democracy, emerging through this negotiated development, and the traditional local democracy of rural societies, in which the local politician is the principal representative.

Yet the problem of these systems is that they are fragile, based on a territorial dimension that may prove to be ephemeral, and their logic is first of all instrumental, as it does not constitute a collective project expressing shared values. Drawing examples from the French agri-environmental experience, Billaud writes that in practice the negotiation process only rarely followed verified criteria. The scope was determined not on the basis of environmental objectives and specifications but on the basis of the available financial resources and their distribution among participants. Given the results of changes, in practice the agri-environmental experience could seem to be in vain.

What has emerged from these programmes is a collective experience of an unprecedented dynamism in the construction of a social relationship to nature. Individuals and social actors built socio-technical 'assemblies' giving official status to social and scientific uncertainty and preaching constant adjustment and compromise as a guide to public action. They offer a form of organization and regulation of social life based on procedural democracy. The rules of the game are derived from extremely localized arrangements and that makes the social agents involved in these forums unable to escape from the collective framework of 'ecological interdependence'. Billaud argues that a possible reconstitution of rural societies depends precisely on the relationship between a technical democracy, proper to forums such as these, and a local democracy, established by virtue of citizenship.

Di Iacovo discusses European agricultural policies and the productive and organizational structures of Italian agriculture, focusing on the possible impact that the new political and socio-economic agenda, including Agenda 2000, could have on Italy. He identifies Agenda 2000 with a move towards a more self-regulated market, a better response to consumer needs and a greater consistency as far as the requirements of local systems of development are concerned.

According to Di Iacovo, these three foundations of Agenda 2000, which are in one way or another conflictual, mirror a new organization of productive relations on a worldwide scale. The globalization of the economy is an uneven process and those benefiting from the redistribution of wealth are not only nation-states but, above all, specific local systems. Consequently, it is not a coincidence that the EU favours principles of subsidiarity and co-financing. The EU seems intent on ensuring a matrix of common references in terms of macro-objectives and instruments, leaving it up to the individual national and local decision-making bodies to adjust their policies according to local trajectories.

With respect to the feature of Italian agriculture two types of agricultural production units are identified. Enterprises that, although organized in different ways, are integrated into international, national and local markets; and organizations whose link with the productive world has waned over time, but remain significant from a social point of view and in terms of land management. On account of structural, organizational, environmental, and territorial diversities of Italian rural areas, the effects of Agenda 2000 must vary from one situation to another.

As far as rural development policies are concerned, the EU points to numerous functions that rural areas can be called upon to carry out in order to requalify development paths. An indirect effect of the CAP with its emphasis on centralized bureaucratic bodies at the national and EU levels is that it eroded the entrepreneurial ability of farmers, and the interest of the political classes and local councils towards problems in agriculture and rural areas. Nowadays, therefore, the start of locally determined development plans implies an intense revision of the interpretation of the categories of 'reciprocity', 'organization', 'market', and 'political exchange'. Negotiation at a local level has once again become important and public administration and the structure of local institutions represent an important key to the local development process.

Demoussis discusses the development of the CAP and summarizes the main reactions to it from the early 1980s onwards, which led to the most recent CAP adjustments and re-evaluations. Turning to Greek agriculture, he underlines the fact that today the CAP contributes to almost half of Greece's total agricultural income. Given the persistent structural problems of the Greek agricultural sector (small and fragmented land holdings, an aging farming population, the bad state of the co-operative movement, the bureaucratic public administration, etc.), it seems only natural that the competitiveness of Greek agriculture has not improved. Contributing to this development is also the fact that Greece, since its admission to the EU, has consistently exhibited its own inability to institutionally intervene with effective national reforms in areas that were not directly in the narrow interests of the CAP.

After a detailed discussion of the major aspects of Greek agriculture the author considers the framework of a dynamic and comprehensive development strategy for the Greek countryside. In such a framework the problem of low competitiveness is dealt with by investments that improve the quality of production, the timely and systematic adoption of new technologies and production techniques, the upgrading of agricultural training and education, and the improvement of marketing strategies.

With regard to rural development, the target should be the maintenance of an economically and demographically viable population in the Greek countryside, particularly where this is more difficult, as in mountainous areas, islands and less developed and border regions. In order to achieve this goal, programmes for integrated and multi-functional development must be implemented. Targeted policies differentiated by region, could lead to substantially smaller inequalities in the Greek countryside. An appropriate institutional framework should be formed to implement such policies. The author suggests that the plan for the development of agriculture in the Third Community Support Framework (Third CSF) (2000 to 2006) is a step in the right direction, because many of the interventions seem to be directed toward what is commonly understood as the integrated development of the countryside.

## References

Barlett, P.F. (1986), 'Part-time Farming: Saving the Farm or Saving the Life-style?', *Rural Sociology*, Vol. 51(3), pp. 289-313.

Bazin, G. and Roux, B. (1995), 'Resistance to Marginalization in Mediterranean Rural Regions', *Sociologia Ruralis*, Vol. 35(3), pp. 335-47.

Boyle, P. and Halfacree, K. (1998), *Migration into Rural Areas: Theories and Issues*, Wiley, New York.

Bradley, T. and Lowe, P. (eds) (1984), *Locality and Rurality: Economy and Society in Rural Regions*, Geo Books, Norwich.

Braudel, F. (1969), *Ecrits sur L' historie*, Flammarion, Paris.
Brouwer, F. and Lowe, P. (eds) (2000), *CAP Regimes and the European Countryside*, CAB International, Wallingford.
Commission of the European Communities (1996), *The Cork Declaration: a Living Countryside*.
Damianos, D. et al. (1998), *Greek Agriculture in a Changing International Environment*, Ashgate, Aldershot.
de Haan, H., Kasimis, B. and Redclift, M. (eds) (1997), *Sustainable Rural Development*, Ashgate, Aldershot.
Drummond, I. and Marsden, T. (1999), *The Condition of Sustainability*, Routledge, London.
Fearne, A. (1997), 'The History and Development of the CAP 1945-1990', in C. Ritson and D. Harvey (eds), *The Common Agricultural Policy*, CAB International, Wallingford.
Fennell, R. (1997), *The Common Agricultural Policy. Continuity and Change*, Clarendon Press, Oxford.
Friedland, W. et al. (eds) (1991), *Towards a New Political Economy of Agriculture*, Westview Press, Boulder.
Friedmann, J. (1987), *Planning in the Public Domain*, Princeton UP, Princeton.
Fuller, A. (1990), 'From Part-time Farming to Pluriactivity: a Decade of Change in Rural Europe', *Journal of Rural Studies*, Vol. 6, pp. 361-371.
Gardner, B. (1996), *European Agriculture*, Routledge, London.
Gasson, R. (1988), *The Economics of Part-time Farming*, Longman, Essex.
Goodman, D. (1999), 'Agro-food Studies in the "Age of Ecology": Nature, Corporeality, Bio-politics', *Sociologia Ruralis*, Vol. 39(1), pp. 17-38.
Goodwin, M. (1998), 'The Governance of Rural Areas: Some Emerging Research Issues and Agendas', *Journal of Rural Studies*, Vol. 14(1), pp. 5-12.
Hadjimichalis, C. and Papamichos, N. (1991), 'Local Development in Southern Europe: Myths and Realities', in E. Bergman et al. (eds), *Regions Reconsidered: Economic Networks, Innovation and Local Development in Industrialized Countries*, Mansell, London.
Kasimis, C. and Papadopoulos, A.G. (1994), 'The Heterogeneity of Greek Family Farming: Emerging Policy Principles', *Sociologia Ruralis*, Vol. 34(2), pp. 206-228.
Kayser, B. (1990), *La Renaissance Rurale*, Ar. Colin, Paris.
King, R. et al. (eds) (1997), *The Mediterranean: Environment and Society*, Arnold, London.
Marsden, T. (1995), 'Beyond Agriculture? Regulating the New Rural Spaces', *Journal of Rural Studies*, Vol. 11(3), pp. 285-296.
Miller, S. (1996), 'Class, Power and Social Construction: Issues of Theory and Application in Thirty Years of Rural Studies', *Sociologia Ruralis*, Vol.36(1), pp. 93-116.
Papadopoulos, A.G. (1999), 'Revisiting the Rural: a Southern Response to European Integration and Globalization' in C. Kasimis and A.G. Papadopoulos (eds), *Local Responses to Global Integration*, Ashgate, Aldershot.
Ray, C. (2000), 'Endogenous Socio-economic Development in the European Union – Issues of Evaluation', *Journal of Rural Studies*, Vol. 16(4), pp. 447-458.
Redclift, M. et al. (eds) (1999), *Agriculture and World Trade Liberalization: Socio-environmental Perspectives on the Common Agricultural Policy*, CAB International, Wallingford.
Rees, P. et al. (eds) (1996), *Population Migration in the European Union*, Wiley, Chichester.
Robinson, G. (1994), *Conflict and Change in the Countryside*, Wiley, Sussex.
Roseland, M. (2000), 'Sustainable Community Development: Integrating Environmental, Economic, and Social Objectives', *Progress in Planning*, Vol. 54(2), pp. 73-132.

Tangermann, S. (1999), *The European Union Perspective on Agricultural Trade Liberalization in the WTO*, University of Guelph, Ontario.

van der Ploeg, J. (1993), 'Rural Sociology and the New Agrarian Question: a Perspective from the Netherlands', *Sociologia Ruralis*, Vol. 32(2), pp. 240-260.

Whitby, M. (ed.) (1996), *The European Environment and CAP Reform*, CAB International, Wallingford.

# PART I

# THE CAP AND THE EUROPEAN COUNTRYSIDE

Chapter 2

# The Condition of Rural Sustainability: Issues in the Governance of Rural Space in Europe

Terry Marsden

## Introduction

To say that the governance of rural space in the European context is at something of a crossroads is, of course, to understate a set of conditions that have been evolving for at least the past decade (Redclift et al., 1999). Despite numerous policy documents and initiatives (like the recent Agenda 2000 associated and rural regulation statements), considerable difficulties persist concerning the development of a truly reformed European approach. For instance, despite over fifteen years of debate and policy crisis concerning the 'arthritic' nature of productionist support mechanisms within the Common Agricultural Policy (CAP), and the need to shift the emphasis towards new social and environmental agenda, it is still the case that the main pillar of the CAP remains in this area in terms of funding. Moreover, it still tends to reinforce the logic of agricultural productivist scale economies by rewarding the largest volume producers, as well as 'locking in' many of the less productive producers and those least able to meet the demands such policy-designed 'technological treadmills' require.

The most contemporary manifestation of this highly asymmetrical support system is associated with the outcomes of the Agenda 2000 reform package and the establishment of what has come to be regarded as the 'Second Pillar' of rural development. Whilst again this recognizes the centrality of a broader sustainable rural development approach to the European countryside and its policies, the amounts of funding in the 'Second Pillar' are likely to remain marginal. For instance, while in 1997-8 England received over 3.625 million pounds in commodity support, representing 25.000 pounds per holding; the forecasts for 'Second Pillar' funding in its first year (2001-2) is estimated as only in the order of 22 million pounds, rising to 70 million by 2006-7. Whilst this does not take into account all the impact of the recent regionally allocated Structural Funding, it nevertheless suggests that however urgent the calls are for a serious reform of public funds away from the productivist logic, and towards a more 'multi-functional' or orchestral approach to rural policy, the inertia in this tendency will require alternative strategies associated with different types of state, community and market-based action. Such inert tendencies at the macro-policy-making level, and in some member-states, represents one feature, I will argue, of a growing contradiction which lies at the heart of current rural development policy discourse. It is a rural development discourse, which to put it mildly, faces a long-term crisis of legitimacy; not necessarily amongst the rural populations of Europe alone, but also amongst the urban consumption classes.

In terms of the growing food consumption concerns, health, welfare, and employment issues, the current system of agricultural and rural regulation in Europe is unsustainable. The question which I will address here is: How is it that these unsustainable conditions can be, in themselves, sustained for so long? And what sort of alternatives are there that at the moment seem to remain marginal but may provide new pathways out of the impasse?

In examining the sustainability of unsustainable conditions it is necessary, I will argue, to explore three social/political/scientific/economic dynamics which are currently combining and competing in shaping European rural space. These are: *the agro-industrial model; the post-productionist model*; and an as of yet emerging and *alternative/sustainable rural development model.* These models of development are all variably insinuated into the current policy and scholarly debates concerning rural development, and particularly the renegotiated role that agriculture could or should play in it. They are in this sense each struggling for primacy in the policy development process. In conclusion, I make some suggestions as to how these may play themselves out with regard to rural Europe. It is critical to recognize how these dynamics do and will have spatial and social effects which in turn condition the degree to which we can move from the foreclosing corners of the unsustainable to the vistas of the sustainable.

The chapter is divided into three sections. First I outline some of the principal rural governance questions which currently confront rural Europe before then detailing the nature of the three dynamics. In conclusion, I speculate on some of the spatial and policy implications of the third model: that of sustainable rural development.

## The Governance of Food and Rural Systems in a Period of Crisis of Legitimacy

The major and continuing pervasive influence upon rural development and particularly the role of agriculture and food within it lies squarely within what can be termed the *agro-industrial dynamic.* As Table 2.1 suggests, this is associated with the globalized production of standardized products and the recent attempts to 'de-regulate' international markets so as to supposedly produce a more level playing field on which agri-business and intensive producers can compete. This is a dynamic which ties agro-food centrally into an industrial dynamic, treats natural food products as industrial products, and tends to see the reform of the CAP as a preface for further concentration in the agro-food sector. This is by far the major development paradigm governing agro-food production in Europe at the present time despite over fifteen years of attempted reform on the basis of reducing its environmental and social effects. Most of the major commodity sectors, particularly in North-western Europe and in parts of Spain and Italy are governed by what I call this 'race to the bottom' paradigm.

*The case of the dairy sector is instructive here* When the Agenda 2000 negotiations started in 1998, the British government gained some supporters (Italy, Denmark, and Sweden to form the so-called London club) for the policy of abolishing milk quotas (by 2006). Although this has been delayed in the most recent (Berlin) summits, price cuts (15 per cent over three years) and relaxation of quotas is the general tendency. This is serving to heighten the need for farm concentration and further reduce mixed farming methods involving dairying.

Globally, this industrialized and globalized competitive model of development is about the survival of the fittest and the further application of intensive technologies which reduce input costs through scaling up. However, as a telling scenario outlined by Agra-Europe suggests, such a process may well be one that the majority of European farmers cannot altogether win:

> The latest figures compiled by a team of international economists on an admittedly small sample of farms in the main dairying areas of the world suggest that even the best European dairy farms would have difficulty competing against the low-cost, grass-based systems of Oceania or even the largest and most efficient dairy units in the western United States. More immediately and importantly, however, these figures suggest that EU producers could not compete with the best central European producers on the basis of current labour and land costs. In addition, the greater potential international competitiveness of the most efficient EU producers, such as those in the UK, could be eroded by currency appreciation (Agra-Europe, 7th August 1998 A, p. 1).

The globalization logic of the *agro-industrial model* is reinforced further by partial deregulation and the somewhat spurious faith in 'free' competition (for instance with abolition of the UK nationally based Milk Marketing Board (MMB) in 1994 on the basis of its monopolistic practices). In the meantime dairying is being practised by larger and more specialized units. In Wales, for instance, after long-term declines in the number of dairy farms in the 1960s and 1970s (down in numbers by 33 per cent between 1964 and 1974); the number of milk producers fell by just over 50 per cent from 1977 to 1996. This downward trend is set to continue.

While 'reform' of the CAP has been slow (either in developing a more effective rural development arm or in more radically disassembling the price and production support structures) it is clear that the realities of the current agro-food conditions are still largely pulling it along an agro-industrial path which is based upon a neo-classical logic of scale and specialization using industrial technologies. Such a model of development, however, is not without its growing threats.

*The Agro-industrial Model: the Management of Unsustainability*

The industrial food governance system, associated with what McMichael and Friedmann, *inter alia*, have called the 'first' and 'second' food regimes, owes its existence to the tendency towards the reduction of distinctiveness of food production, supply and accumulation, and the alliance between food and other technological trajectories associated with industrial activity – where there is an in-built requirement to overcome the vagaries of nature and a rational approach towards increases in outputs. This, of course, has been the dominant model of food provision in the post-war period (Buttel, 1997), and has been conditioned in ways which attempt to stave off the unacceptable environmental consequences of such a system by developing new technologies (e.g., biotech and Genetically Modified Organisms (GMOs)).

However, the direction and development of this model has not been without its tensions over the past decade. It is worth illuminating these on the basis of identifying how the model has attempted to 'solve' – perhaps only over the short or medium term – some of its own problems. For instance, over time agri-businesses and corporate retailers have not only had to cope with the intensity of competition amongst themselves; they have also had to continually innovate in ways which at least 'seem' to solve the growing consumption and environmental concerns associated with the foods they have provided. This is a sort of competitive and consumption dynamic which has steadily come to grip these fractions of food capital. As this grip has tightened, the conventional techniques of the state have been found wanting, so that private capital can no longer rely upon the productivist 'state guarantee'.

Nevertheless, this situation does not immediately stop the state from trying to exercise governance. Various governance mechanisms have been incorporated under this model; many of them from different origins. Many are now undergoing considerable crisis – all at the same time. The synergies between the different vectors of regulation have been subject to crisis. What are these vectors and why are they experiencing crises?

**The Crisis of State-supported Standardized Food Production**

The CAP in Europe, while continuing to represent the main plank of agro-food policy, holds increasing contradictions. While its aims have never been to foster the supra-productionist goals of large-scale farmers and agro-industries alone, major allocations of funding have favoured the most productive producers, meaning that where payments have been made to the more 'marginal agricultural sectors' (e.g., hill farmers, small producers in Southern Europe) the support has been less and less sustainable either in an economic or environmental sense. The recent policy reforms under Agenda 2000, in addition to the new rural development regulation, expose a policy framework which will do little to shift the basic philosophy beyond its bias towards the industrial model. Under Agenda 2000, more conditional payments to marginal farmers will be more spatially specific and dependent upon the national enthusiasm to matching fund incentives. In countries like the UK, the overall emphasis looks as if it will still be geared towards the pressure for the most productive farmers to meet world market prices rather than uphold conditional welfare payments to the many smaller farmers who can no longer reduce their costs through economies of scale. However, it is clear that change in the CAP regime will come, if only slowly. This period will be critical for the development of alternatives in food production and supply. The question will be how long farmers will manage to stay upon the CAP-induced technological treadmill. How many will wish to take a 'jump in the dark' in another direction? And how many will be able to?

Inside the more post-productivist CAP, then, we see an inherent contradiction which can only widen as the pragmatic and more national-based journey of the CAP reform unfolds. On the one hand we see the more welfarist steps which tend to embrace a rural development paradigm (most notably reflected in the discussions concerning the new 'rural development regulation'). This is a policy which goes some way beyond the conventional image of equating agriculture with food. On the other hand, there is a policy which has the industrial innovation system as its supporter; the super-productionist policy of encouraging those producers who can compete on the 'world market' to do so using whatever technologies (including GMOs) and Research and Development (R&D) systems necessary. Increasingly, this arm of the CAP, and its wider relatives, such as the World Trade Organization (WTO) and structural adjustment, etc., becomes less and less associated with the former paradigm in that it is associated more with the logics of industrial commodity markets (for instance, the use of food products for industrial products such as paints and energy) and an innovation system which continues to reduce human and natural inputs into the production and processing stages.

This inherent contradiction of the state-supported system has the effect, in part because of the longevity of the transitions occurring, of creating radical dissonance in the structure and functioning of the local and regional production systems. In short, as in the case of the classic bell curve, only the minority of producers are clear about where they are going in this system: the productionists who are of the scale and integration to go for world markets; and the 'multi-functional farmers' now individually and locationally capable of developing a sustained alternative and hoping to capture the new, conditional welfare payments the revised CAP might bring.

In regulatory terms then we can see that the evolution of new alternative food governance systems is not as simple as a binary contradiction between the industrial and the 'alternative' food systems; or that we are entering, somewhat unproblematically, a rather more complex 'third food regime'. Rather, we can see the conventional systems of policy 'struggling with themselves' as they try to accommodate more external and political demands.

## The Rationalist and Industrialized Response to the Quality Food Crisis

A second dimension of this industrial dynamic concerns the ways it is responding to what we might call the 'quality food crisis'. The subsidization of a food production system which has encouraged cheap inputs, reductions in labour, and high levels of mechanization is clearly faltering. In its place, but again somewhat variably, are now placed a whole array of regulatory measures (designed and implemented by the government and the private sector) to ensure food safety and hygiene. These have tried to 'tidy up' the industrial food chain at the edges (for instance, by cutting even more fat off meats and vacuum packing uniform weights of meat), and represent an industrialized and rational response to the quality food crisis (EC, 2000). As a result, the bulk of the regulatory costs of such 'food hygiene' techniques have been placed either upon the producer or the consumer. In addition, the imposition of such regulatory rules operate as important criteria for market entry, not only for the producer, but also for the food processing company struggling to keep contracts with the multiple retailers. In this context, the provision of 'new' quality lines in the superstores (e.g., 'freedom foods', welfare lines, free range, etc.) really only represent new innovations in supermarket category management principles (Cook and Crang, 1996), rather than clearly defined alternative food supply networks. They do little to emancipate either nature, region or producers in the supply chain; and they supplement the appropriation of the commodity with the appropriation of the quality conventions upon them. Controlling these quality conventions becomes a key factor in the maintenance of the corporate-led system which now continues to innovate through rational means to provide sanitised products of low risk but of higher value.

## The Marginalization of Alternatives

Despite the variations and tensions which exist in this industrial model, its principles remain largely intact. Nature, region and quality are configured in ways which suppress variation in the standardized lines of food supply. Innovation and capital investment are located towards the retail end of supply chains; and, above all, food prices, while perhaps more variable, are kept under control. This latter factor (assisted of course, through the state-supported CAP, e.g., export levies) means that alternatives to the industrial system can be marginalized on the basis of price. Thus a major effect, intended or otherwise, of the rationalization of food supply chains and the continued innovation of technologies which make food 'more safe' (if not of more nutritional quality), is to make it all the more difficult for alternative food supply chains to experience a 'take-off'.

Nevertheless, as is the case with the GMO issue, consumers are increasingly conscious of the composite problem with industrialized food supply 'from plough to plate'. The diversity of concerns, from animal welfare, environmental conditions, ethical questions, food quality and health do not act separately, but they combine to counter the marginalization effects of alternatives. This creates a major problem for the industrial system, which bases its innovations in food quality on the specific commodities themselves (i.e., rationalizing and purifying foods in ways which enhance food safety and provide various grades of food quality as a form of value added). It is increasingly difficult for this system to deal with the *composite nature of food consumer concerns despite high levels of investment and innovation.* The struggle, therefore, develops between the continued marginalization effects of industrially produced 'cheap foods' on the one hand, and the evolving composite consumer effects on the other which expose the hidden costs in the industrial system.

These different dynamics play a key role in destabilizing any form of hegemony in overall 'food governance'. European Union (EU) and national government policy cannot be consistent in dealing with these dynamics, and this fuels growing legitimation problems both for itself and for the industrial model. Alternative food politics begins to fill the gaps left by legitimate government regulation, systemic questions about the provenance and manipulation of foods (Goodman, 1999). However, retailer-led supply chains based upon rational and self-regulated systems of quality control continue to dominate food supply (Marsden et al., 2000) and the primary producer sector becomes increasingly dependent upon such systems of market/quality control. Value is increasingly abstracted from foods at the consumer end of the supply chain as part of the imposition of these quality conventions. This tends to further empower the large retailer. Government and the private sector have reacted conservatively to the growing crisis of legitimacy, opting to keep in place the basic principles of the industrial system while at the same time highlighting a rational conception of food quality.

## The Post-productivist Dynamic

Despite the salience of the agro-industrial development path, the past decade has also witnessed the growth of a new 'post-productivist' dynamic which has challenged the relevance of industrialized agricultural production in many rural areas of Europe. Under this model, most easily applied to the rich counter-urbanized regions of north-west Europe, many scholars and policy-makers now deny the social and economic significance of agricultural production given its relatively low contribution to Gross National Product (GNP), regional levels of employment, and the increasing dominance of public and private sector services. The combination of the decentralization of industry and services and the relentless stepwise outflowing of the richer urban residential population (Boyle and Halfacree, 1998), not only around the larger metropolitan areas (such as London and Paris), but also deeper into the urban and rural fabric, has turned much of the receiving countryside into 'consumption spaces' (Marsden et al., 1993; Marsden, 1999).

Such a perspective is now largely taken for granted by many scholars and policy makers in Northern Europe. And it is reinforced by the recognition that the problems of rural development, of poverty and social exclusion, have not and cannot be solved by a focus upon the agricultural alone. Rather, it is the provision of rural services (both public and private) which can hold a significant key to this problem. The growth of the 'consumption countryside' model also holds a particular conception of nature and the rural. While, for instance, the 'agro-industrial model' sees rural nature as something to be overcome, or at least held back by the continuous application of new technologies in the agricultural and food sphere, the post-productivist model tends to marginalize nature in a different way. Here, rural nature is now a consumption good to be exploited not by industrial capital but by the urban population (Enticott, 2000). Moreover, it is usually a specific form of nature that needs to be constructed and exploited. It is centrally concerned with the construction of the rural landscape and its protection. Such a project is seen as somewhat disembodied from agricultural developments, with producers now being seen as the potential 'criminals' in the moral crusade to protect these cherished rural spaces (Lowe et al., 1997).

Underlying the 'post-productivist' conception of the countryside then is a pervasive 'moral economy' with respect to agriculture. The latter is seen as a 'dirty business' to be cleaned up by progressive environmental legislation. In the meantime a raft of non-agricultural policies can be more effectively directed at maintaining a rural economy and generating 'thriving rural places'. Such conceptions of the countryside are thus not just associated with the growth of the ex-urban middle class as the power-brokers in the countryside, but also associated with a particular conception of rural nature; one which places most emphasis upon the protection of the visible landscape and the maintenance of the scarcity value of different types of rural environment.

So, while the *agro-industrial model* tends to marginalize rural nature through the production process, *the post-productivist model* now does the same thing through a highly materialist conception of the consumption process. Not coincidentally, both sets of modernist conceptions and models stem from assumptions about *external controls over rural nature*, and they also involve the development of external forms of governance over the rural in ways which attempt to position nature so as to be at least sustainable for some period of time in the future. Such conceptions are now embodied in government reports. For example, a recent Cabinet Office report in the UK argues from the outset that:

> Rising affluence and changing tastes have led to a reduced emphasis on the countryside as a resource for production and a greater emphasis on its potential for 'consumption', both as a place in which to live and in which to enjoy leisure and recreational opportunities.... More people see themselves as having a stake in the countryside (and in particular, in the state of its environment): and the views of rural residents have diversified, reflecting the changing social composition of the rural population (Cabinet Office, 1999, p. 29).

The consumption countryside is, however, a site of conflict. Continuing with the dairy sector as our reference point, we can see under this conception of development how this intensively-organized sector contradicts the growing moral economy of post-productionism. This is most clearly demonstrated in Lowe et al.'s study (1997) of dairy farmers and pollution in Devon, England. As they conclude:

The politicization of farm pollution led to the imposition of various controls on agriculture to restrain the excesses of agricultural productivism. What these environmental regulations meant in practice was that farmers were increasingly confronted by regulatory officials armed not only with new powers but also with a new moral authority. The exchanges that took place between them were not limited to what constituted sound agricultural practice, but touched on nature, morality and the law. Much of what was of interest to us not only in the regulation of polluting agricultural practices, but also in terms of the making and undoing of moral discourses, occurred in these encounters. What was revealed was a traditional order (of agricultural productivism) under challenge and a new order (around environmentally responsible agriculture) being discursively constructed and resisted (Lowe et al., 1997, p. 207).

What we see here is the conflict between productivist and post-productivist models being played out between the farm and wider rural population. A population that now has to confront not only the vast regulatory system associated with the agro-industrial model (as outlined above) but also the highly diverse and moral regulatory system stemming from a particular form of (externally imposed) environmental care. This form of environmental care or social management tends to criminalize dairy producers while protecting and shaping a countryside in ways which will remain particularly attractive to the maintenance of the counter-urbanization process. Thus the *post-productivist* countryside model of development is founded on anything but environmental protection 'for its own sake.' It is far more functional than this. It is about socially and morally shaping the countryside in ways which continue to make it attractive to aspiring ex-urban social elites.

There are, of course, at the time of writing, several types of rural backlash to these trends. In the UK the Countryside Alliance movement has grown rapidly into a force for projecting the anti-urban view of the countryside, and in the US the anti-environmental movements (Jacobs, 1999) project both productivist and various types of stewardship perspectives for landowners and farmers. These conflicts occupy one contested territory concerning rural nature; a territory which brings together productivist with consumption concerns. Both, however, tend to marginalize and deny the critical role of sustainable agriculture in socially managing rural nature.

*The Sustainable Rural Development Dynamic: Pathways Out of Contradiction?*

Both of the aforementioned rural development dynamics have their own socio-spatial expressions. In many rural regions in Europe they overlap across rural space and affect change in two ways. For instance, the quiet progression of the industrial agro-food model (for instance in the intensively farmed lowland regions) tends in many rural spaces to continue as the post-productivist conditions also take hold. This creates more complexity about the ways in which rural nature is being governed; for instance, externally managed CAP funds and quality conscious corporate retailers condition the actions of smaller groups of specialized producers, while at the same time local and regional environmental protection and planning attempt to contain and shape the same rural landscape.

The interesting contrast here is that the former (productivist) system of governance tends to stress the national and international perspective (largely ignoring local and regional specificity of circumstances). The latter, by contrast, governs by local means and places giving strong emphasis upon the specification of the local environment.

Both, however, tend to downplay the potentially symbiotic social and environmental roles of agriculture. As Kropotkin foresaw 'land is going out of culture at a perilous rate', implying the growing severance between *ager* and *cultura*-field from farming community, and nature from human society. Bookchin also perceived this trend, and urged a radical alternative which would be an extension of early farming practices, belonging to a 'moral economy' that views land as an *oikos* of living bacteria, insects, plants and animals: such an alternative would be in direct contrast to the onset of *agribusiness* (which belonged to a 'market economy' which treats the earth as a resource to be exploited, denying its cultural significance (Macauley, 1999).

One important lesson concerning the onset and dominance of both the agro-industrial model and the post-productivist model is that their own specific moral economies, there own specific ways in which social nature is managed, rely upon both market and state governance structures which attempt over the medium term to manage the largely unsustainable conditions which they create (Drummond and Marsden, 1999; Drummond et al., 2000). These governance structures, whether associated with agro-industrial productivism, or differentiated types of locally based post-productivism, maintain a tight grip in their attempts to avoid environmental and social crises of legitimacy. As Drummond et al. (2000) indicate when referring to the growing crisis of European agriculture:

> Thus far, both the analysis and policy have failed to relate the progressive *devaluation of nature* to the reconstruction of regressive modes of social regulation, and debates about sectoral policy have progressed as if these interconnections and dialectics are inconsequential. It has generally been assumed that environmental and social policy can be bolted onto both agricultural policy and, more importantly, onto existing structures and relationships'. On the one hand policy is being articulated in terms of environmental and social goals, but because these are to be implemented through the existing farm structure, this may well serve to reinforce and reproduce the tendencies towards continued instability and environmental instability...(Drummond et al., 1999, p. 124) (my emphasis).

These regulatory and governance mechanisms have a dual role in both upholding largely unsustainable agricultural practices and broader rural conditions and acting as *significant barriers* toward more radical overhauling of rural policy. In short, the productivist and post-productivist systems of governance seem to be able to co-exist, at least for the time being.

Perhaps it is not surprising, then, that those regions which have been *least exploited by either model* which have given more initial impetus to the alternative *sustainable rural development model*. While it may owe its origins to those regions which have been largely 'passed over' by the other development logics, it is now highly pertinent to their continued sustainability (Table 2.1).

What are some of this model's emerging features in relation to agro-food? To what extent can we begin to see alternative and robust regulatory and governance structures taking hold? Quite significantly we see a broadening definition of nature, which now encompasses not just its surface or landscape value as a consumption good, but one which gives more emphasis to food production and the natural transformation process in itself. In addition, the social and economic aspects of sustainable development are being brought more clearly to the fore. So to what extent does this represent a new departure for rural governance of nature?

In variable ways this alternative rural development model conforms more effectively to Guzman and Woodgate's (1999) definition of agro-ecology which encompasses:

> the ecological management of biological systems through collective forms of social action, which pursue co-evolutionary pathways that address the needs of society/nature without jeopardizing the integrity of society/nature. This is to be achieved by systemic strategies that promote the development of more equitable and flexible forces and relations of production, and modes of consumption and distribution. Central to such strategies is the local dimension where we encounter endogenous potential encoded within landscapes and knowledge systems (local, peasant or indigenous) that demonstrate and promote both ecological and cultural diversity. Such diversity should form the starting point for alternative agriculture and for the establishment of dynamic yet sustainable rural societies (Guzman and Woodgate, 1999, p. 26).

While this is an over-arching and ideal-typical definition, there is now sufficient evidence across Europe that many of these components are beginning to emerge in contrast to our earlier and more dominant models, at a regional, rather than at a national or specifically local level (Marsden et al., 1999).

Such a definition is marked by a different set of organizing principles which place nature, labour, region, value, and quality in a different set of equations. Not only are the foods themselves representative of different quality conventions (i.e., ecological, spatial, artisanal) but the mode of social organization of supply chains is radically reorganized. The connectedness of supply chains is related to new types of association, and the types of innovation are linked much more to organizational as opposed to technical developments. The factors of production are reconfigured in ways which place capital in a more symmetrical relationship with labour and land, and emphasis is placed upon the creation of positive externalities from a more multi-functional land-based production base. The emphasis is upon diversity and divergence rather than convergence and consistency. Production and supply chain forms are more highly variable and 'context-dependent'. Value is captured more effectively at the producer end of the chain and the use of spatial identifiers for the foods produced becomes an important dimension (Table 2.1).

Such an 'alternative system' requires new forms of quality regulation to legitimize its presence. Outside the EU-designated categories (Protected Denomination of Origin (PDO's), etc.) and various certification bodies associated with organics and green products, these tend to (i) be more recent, (ii) be locally defined on the basis of the production unit or the locality; and (iii) are more associated with satisfying the composite character of the consumer concerns mentioned above. These regulatory systems have not simply developed 'top down'; and national government policy and regulation is currently only marginal to the governance of the system.  Indeed, because of the diversity and the emphasis upon place, nature and quality, national governments find it increasingly difficult to control and regulate the alternative system. This is one of its main features, and it represents a major constraint upon the widespread development of these systems. A major reason why the rational and industrial systems of regulation are often seen as more attractive and legitimate from a government point of view is that they are more 'controllable' and therefore more conducive for adoption, particularly under conditions of growing crisis.

In recent years, a number of alternative approaches to food governance have emerged, their development coinciding to the rapid growth in influence and power of the multiple retailers. The emergence of 'forces of opposition' to the dominant model of food provision raises interesting questions, especially at a time when the hegemony of governmental influence over food policy is in crisis. The Bovine Spongiform Encephalopathy (BSE) crisis, and to a lesser extent, the debate over GMO foods have done much to undermine the public view of the legitimacy and impartiality of the state in situations where agro-industrial and consumer interests collide (ESRC, 1999; Marsden et al., 2000).

Against this new background, new forms of food governance have been emerging on an ad hoc and grassroots basis. In the past five or so years they have acquired a 'critical mass', challenging both commercially and ideologically the dominant modes of governance. What are some of the patterns of development in these alternative food networks?

**Product-focused Alternative Food**

Of the new modes of food governance being devised at present, the most significant, for a number of reasons, is that of organic produce. Although not 'new,' the conscious adoption of organic standards by increasing numbers of farmers and consumers represents a potentially significant challenge to the industrialization and intensification of agro-food systems, reflecting a desire to return to more natural values. This would not have come about had it not been for the growing concerns over the safety, health and environmental and animal welfare implications of the existing system, all of which are implicitly challenged by the rigorous standards (now defined by EU legislation (CEC 2092/91) but open to slight variations in interpretation at the national level) which ban the use of artificial fertilisers, pesticides and pay careful attention to animal welfare. Consumer demand for such products continues to rise and far outstrips supply, – especially in Northern Europe (Micheslen et al., 1999).

While the emergence and growth of thriving organic sectors across Europe poses challenges to the dominant system of food governance, it also offers new opportunities. On the production side it implies withdrawal from the agrochemical complex, greater reliance on local sources of fertility and a reversal of the fifty-year process of intensification and specialization. Yet it also offers opportunities to tap into new markets at a time when commodity prices, for example of milk, have sunk to a fifteen-year low. This is especially relevant for marginal regions in both Northern and Southern Europe, which cannot realistically expect to compete with the new super-productivist, global regimes brought about by the new CAP and WTO agreements.

After more than a decade of stop/start policies towards organic produce, the multiple retailers are showing signs of serious commitment to promoting and developing the sector. However, engaging with the organic market is not entirely a risk-free strategy for the multiple retailers. Namely, with organic produce more widely available on the shelves of supermarkets, consumers may be given more reason to question the provenance of 'non-organic' produce. As a recent report from the (Danish) Ecology Board points out that 'in today's environmental reality consumers demand organic milk in order to avoid milk which contains pesticides. In the old days this was simply called milk. And so it should be again. Until then the alternative should be labelled pesticide milk. Just as supermarket shelves should contain eggs and battery eggs' (cited in Halkier, 1999).

## Regional, Local, and Artisanal Foodstuffs

A parallel trend in production and consumption patterns is the emergence of a strong and growing market for regional, local and artisanal foodstuffs. This represents a major discontinuity from the assumed dominant trend within the food industry for centralization, standardization and increased homogeneity. In many respects such products represent a manifestation of the 'small is beautiful' movement. Their defining characteristics are small scale of production, using high-grade raw materials (often but not necessarily of local provenance) and artisanal/traditional production techniques to create quality items which attract premium prices which for the most part are sold through local and/or specialist shops rather than large retailers.

Interest in such products stems from a number of sources including the rapid growth of a revised 'food culture' and increasing concerns over the safety of mass-produced foodstuffs (Nygard and Storstad, 1998). The growth of the domestic market for such products is rapidly growing in the UK; according to one survey by 20 per cent over the past three years, and now accounting for 5 per cent of grocery expenditure (DTZ Pidea Consulting, 1999).

Some products are genuinely rooted in tradition (e.g. Scotch Oatmeal cakes) while others may have long traditions of production but have only been recently rebranded themselves in order to differentiate themselves from similar products (e.g. Welsh Black Beef), and some may be relatively new product lines which, nonetheless embody quality and authenticity. Unlike their equivalents in Southern Europe, few British artisanal foodstuffs have, as yet, been accredited under the Protected Denomination of Origin (PDO) and Protected Geographical Indication (PGI) system.

In Ireland, an interesting battle has developed between the *agro-industrial model* (as applied to dairying) and alternative rural development model solutions.

The latter has been adopted in such schemes as the County Clare Partnership Dairy Action Research programme, designed to add value and to diversify the dairy product base for the smaller dairy farmers (Kinsella and Mannion, 2000). Also, as McDonagh and Commins (1999) indicate, the growth of what they term the 'speciality sector' is part of the Irish rural renaissance, even if it is not completely clear whether this will lead to dual strategies of agro-industrial incorporation or continued independent adaptation. Nevertheless, what we witness here in these examples from the Irish dairy sector (as well as some examples in the Dutch dairy sector) (Van der Ploeg, 1999) are counter-productivity strategies as part of a greater rural development paradigm. These focus upon 'economies of scope' (Saccomandi, 1997) rather than economies of scale. The case of dairying as an exemplar here (as followed through with examples from the agro-industrial, post-productivist and rural development models alone) demonstrates the different social management and governance of a natural rural product. The examples also demonstrate that the dominance of the agro-industrial 'race to the bottom' scenario for the dairy producer sector is by no means inevitable; it just has to be socially managed in ways which heighten the significance of milk products, the land base and the producers as social/ natural actors in food governance systems.

Interestingly, these regionally identified products have emerged at a time when there is an increased emphasis on the broader cultural assertion of regional and local identities. They also emerge at a time of economic crisis in farming, when the mantra 'diversify diversify' is central to many governments' thinking on the future of agriculture. It is recognized that 90 per cent of such companies are rural (Ministry of Agriculture, Food and Fisheries, 1999) in the UK, and play a growing role in the economy of the countryside. Their future growth and the propagation of new businesses are likely to be key planks in the forthcoming discussions about the future of the regional rural economy.

**Direct Food Supply Chains**

One important aspect of the centralization and standardization of the industrial food command system developed by the multiple retailers include requirements for continuous supplies of high volumes of standardized produce. These requirements have effectively excluded artisanal, regional and speciality foodstuffs, and in many cases organic produce from the dominant food chain.

In reaction to the difficulties faced by small-scale producers in meeting standards (whether of quantity or quality), a number of new initiatives based on developing more *direct, often locally-based supply chains* are emerging. These include: box delivering schemes, developing closer links to local retail and catering trades, and farmers markets; the use of the internet and mail order (Mai and Ness, 1998), farm shop sales, and at the same time a suspected increase in rural areas in the black market of farmgate meat sales. There has also been a marked increase (after some fifty years of decline) in urban food production, with a growth of interest in the most direct form of a food supply chain: self-provisioning (Martin and Marsden, 1999).

We argue that the development of these new 'short supply chains' represents a more profound shift in governance patterns than the emergence of new products, in that they are being explicitly used to create markets independently of the industrialized and intensive ones led by the globalized multiple retailers. Markets are being forged with new forms of social and association of relationships, where alternate quality standards, generally favouring the ecological and the artisanal come to the fore, where there is a reassertion of the value of the local, and whereby producers, or those closely linked to them, retain a greater proportion of the value added. Many of these initiatives have been led by organic producers frustrated at the difficulties of working within the dominant food supply system.

In the development of these alternative types of food networks, the relationship between food production and distribution on the one hand, and environmental sustainability, human health and regional development are made explicit, with producers, retailers and consumers apparently aware of a broader range of values other than the quality (defined in cosmetic terms) or the cost of food itself. The provision and consumption of food provides a link to issues encompassing health, nutrition, the environment, region, community and third world development. The divide between production and consumption is bridged by direct producer-consumer contacts and the renewed seasonality of produce, now virtually abolished by continuous availability of most lines in the agro-industrial system.

## Policy and Spatial Dimensions: Moving from the Unsustainable to the Sustainable

This chapter has examined three ideal types, all of which are pervasive dynamics shaping the governance of nature and European rural space in contested ways. They stem from different origins and they owe their existence to different sets of social, economic, political and academic/scientific factors. The *agro-industrial model*, for instance, is upheld and reinforced by agricultural economists; many of which are still bemused by the scale and technological expressions imported from the American agricultural modernization project. The *post-productivist project* is associated with social geographers' and planners' concerns about rural capacity and the coping mechanisms needed for dealing with the 'consumption countryside'.

*The rural development dynamic*, while originating in the largely 'bottom up' initiatives associated for integrating and empowering 'peripheral' rural communities, (most notably the LEADER initiatives) is now a much broader and diffuse doctrine; one which can incorporate 'dusted down' ideas of former agricultural practices and social ecology (Light, 1999). All three of these are, in themselves, types of modernization projects. The question becomes which will come to predominate and how do they play themselves out over the differentiated rural space?

The exploration of these three dynamics, and particularly the emerging but so far marginal rural sustainable development dynamic, raises some important questions concerning the *regulation and governance of nature and rural space*.

The rural development model demands, albeit in variable form, that nature, region, quality and value be reconfigured around the character and shape of rural resources (Guzman and Woodgate, 1999). Of course, as the post-productivist model reveals, rural areas are more than the sum of their agricultural parts, but we now see pathways for bringing agriculture 'back in' as a significant manager of rural nature.

However optimistic we may be, the renewed certainties such a rural development model may bring, or indeed, excellent exemplars it already portrays (Marsden et al., 1999), should be seen in light of the fact that its very development confronts many of the barriers associated with the established governance and regulatory systems of the other development models. These are clearly in evidence in the policy rhetoric from the EU and various member states. At one level, for instance, the rural development initiatives associated with the CAP reform would suggest whole-hearted support for the third model. But its fledgling birth coupled with its differential adoption by member states demonstrates the difficulties such a model has in competing with the logics of agro-industrialism and post-productivism. Everybody agrees that the CAP should be reformed, but many member states (including the UK) are tending to hide behind a contradiction. Namely, reform means movement towards both the globalized agro-industrial model and the rural development model (through, for instance, the initial development of the 'Second Pillar').

What is important to recognize in this seeming 'reform' is the maintenance of the inherent conflict between the agro-industrial model and the rural development model. Most governments will give their blessing to the development of 'value-adding quality food production'; but they do so with their other hand firmly shaking the super-productivist hands of global agri-business, and condoning all that this implies for the continuance of the productivist agricultural treadmill. This can only block sustainable rural development through marginalizing its impacts.

In conclusion, therefore, in governance and regulatory terms we cannot expect, it would seem, national governments or corporate firms to fully stimulate the rural development model. *Its impetus will have to come from building alliances from within – at least at the regional and local levels.* The governance of rural space, however, is likely to become even more – however reluctantly – drawn into the redefinition and valuation in nature. This will take it far beyond the simple exploitative *modus operandi* of the agro-industrial or post-productivist models of rural development. The big question is how can emerging and durable regulatory and governance frameworks be developed so that existing sustainable rural development exemplars turn into common trends and practices across rural Europe?

We have seen an agro-food industrial system which is prone to increasing crisis; while an emerging and more sustainable rural development paradigm is still held back in its wake. The real culprits of such an unsustainable environmental and social impasse are those who preside over its governance, both in the private and public spheres. The development of a new agro-food governance, one which would play a fuller part in assisting the achievement of sustaining rural nature, becomes more and more urgent.

**Table 2.1 The three competing rural development dynamics**

| The agro-industrial dynamic | The post-productivist dynamic | The rural development dynamic |
| --- | --- | --- |
| Standardized products | Rural space as consumption space | Integration |
| Capital intensity | | Re-embedded food supply chains |
| | The marginalization of agriculture: declining industry | |
| Optimum (quantitative level) of production | | New policy support structures |
| Long/complex supply chains | Agriculture's share of national income falls from 2.9 per cent in 1970 to 1.0 per cent in 1998 | Associational designs and networks |
| High levels of public funding | | Revised combinations of nature/value/region and quality |
| | Rural land as a development space | |
| Continual development of 'technological fixes' | | |
| | Social exclusion | Rural development as counter movements |
| Decreasing value of primary produce and production structures | Public sector services | Rural livelihoods/fields of activity/new institutional arrangements |
| | The social economy and the use of the natural as an attractor in the counter-urbanization process | |
| Economies of scale | | Agro-ecological research and development |
| Rural space as agricultural space | | Co-evolving supply chains |
| Private-interest regulation (led by retailers)/public interest regulation: crisis management and nature management | | Revised state/market/civil society/nature relations |
| | | Evaluation paradigm for rural sustainability |

**References**

Agra-Europe (1998), *Labour and Land Costs Handicap EU Dairy Farms.*
Boyle, P. and Halfacree, K. (1998), *Migration into Rural Areas: Theories and Issues*, Wiley New York.

Buttel, F. (1997), 'Some Observations on Agro-food Change and the Future of Agricultural Sustainability Movements', in D. Goodman and M. Watts (eds), *Globalizing Food: Agrarian Questions and Global Restructuring*, Routledge, London.

Cabinet Office (UK) (1999), *Rural Economies: a Performance and Innovation Unit Report*, London.

Cook, I. and Crang, P. (1996), 'The World on a Plate: Culinary Culture, Displacement and Geographical Knowledges', *Journal of Material Culture*, Vol. 1(2), pp. 131-153.

Drummond, I. and Marsden, T. (1999), *The Condition of Sustainability*, Routledge, London.

Drummond, I. et al. (2000), 'Contingent and Structural Crisis in British Agriculture', *Sociologia Ruralis*, Vol. 40(1), pp. 111-128.

DTZ Pidea Consulting, 1999.

Economic and Social Research Council (1999), *The Politics of GM Food: Risk, Science and Public Trust*, Special Briefing No. 5. Swindon.

Enticott, G. (2000), 'Heterogeneous Ruralities: the Place of Nature and Community in the Differentiated Countryside', Unpublished PhD thesis, Cardiff University, Cardiff.

European Commission (1999), *White Paper on Food Safety*, CEC, Brussels COM, 719 final.

Freidmann, H. and McMichael, P. (1989), 'Agriculture and the State System: the Rise and Fall of National Agricultures, 1870 to the Present', *Sociologia Ruralis,* Vol. 29, pp. 93-118.

Goodman, D. (1999), 'Agro-food Studies in the "Age of Ecology": Nature, Corporeality, Bio-politics', *Sociologia Ruralis*, Vol. 39(1), pp. 17-38.

Guzman, E. and Woodgate, G. (1999), 'Alternative Food and Agriculture Networks: an Agro-ecological Perspective on the Responses to Economic Globalization and the New Agrarian Question', EU COST workshop paper (unpublished).

Halkier, B. (1999), 'Consequences of the Politicization of Consumption: the Example of Environmentally Friendly Consumption Practices', *Journal of Environmental Policy and Planning*, Vol. 1(1). pp. 25-41.

Jacobs, H. (1999), *Land Tenure Centre Newsletter*, Department of Rural Sociology, University of Wisconsin, Madison.

Kinsella, J. et al. (1999), *Creating a New Future for Dairy Farm Households: a Key Element of Rural Renewal*, Department of Agribusiness, Extension and Rural Development, Faculty of Agriculture, University College Dublin, Dublin.

Lekakis, J.N (ed.) (1998), *Freer Trade, Sustainability, and the Primary Production Sector in the Southern EU: Unraveling the Evidence from Greece*, Kluwer Academic Publishers, Dordrecht.

Light, A. (ed.) (1999), *Social Ecology after Bookchin*, The Guilford Press, New York.

Lowe, P. et al. (1997), *Moralizing the Environment: Countryside Change, Farming and Pollution*, UCL Press, London.

Macauley, D. (1999), 'Evolution and Revolution: the Ecological Anarchism of Kropotkin and Bookchin' in A. Light (ed.), *Social Ecology after Bookchin*, The Guilford Press, New York.

Marsden, T.K. (1999), 'Rural Futures: the Consumption Countryside and its Regulation', *Sociologia Ruralis*, Vol. 39(4), pp. 501-521.

Marsden, T.K. et al. (1999), 'Sustainable Agriculture, Food Supply Chains and Regional Development', Special Issue of *International Planning Studies*, Vol. 4 (3).

Marsden, T.K. et al. (2000), *Consuming Interests: the Social Provision of Foods*, UCL Press, London.

Marsden, T.K., et al. (1993), *Constructing the Countryside*, Boulder, co: Westview Press, London.

Martin, R. and Marsden, T.K. (1999), 'Food for Urban Spaces: the Development of Urban Food Production in England and Wales', *International Planning Studies*, Vol. 4(3), pp. 389-413.

McDonagh, P. and Commins, P. (1999), 'Food Chains, Small-scale Enterprises and Rural Development: Illustrations from Ireland', *International Planning Studies*, Vol. 4(3), pp. 349-373.

Micheslen, J. et al. (1999), 'Dissemination of Organic Farming in 18 European Countries: a Description of Variation in National Response to International Regulation', Conference paper, European Society for Rural Sociology, Lund.

Ministry of Agriculture, Fisheries and Food (1999), *The Future of Agriculture in England*, London.

Nygard, B. and Storstad, O. (1999), 'De-globalization of Food Markets? Consumer Perceptions of Safe Food: the Case of Norway', *Sociologia Ruralis*, Vol. 38, pp. 35-53.

Redclift, M. et al. (eds) (1999), *Agriculture and World Trade Liberalization: Socio-environmental Perspectives on the Common Agricultural Policy*, CAB International, Wallingford.

Saccomandi, V. (1998), *Agricultural Market Economics. A Neo-institutional Analysis of the Exchange, Circulation and Distribution of Agricultural Products*, Van Gorcum, Assen.

# Chapter 3

# Mediterranean Agriculture in the Light and Shadow of the CAP

## Nikos Beopoulos

In the thirty-five years of its existence, the Common Agricultural Policy (CAP) has contributed decisively in the transformation of agriculture for member states by means of a common rural development philosophy, common institutions, and an extensive set of policy measures. It is often found, however, that not all transformations, completed or under completion, promote themselves at the same rate. Despite the common elements among farming systems in the different parts of Europe, the process of agricultural integration remains un-integrated among farming systems that are markedly coloured by their geographical and historical heritage.[1] Mediterranean Europe's bio-climatic uniqueness and age-old history (imprinted on its rural landscapes and its cultural identities), for example, set it apart.

The existence of geographical regions such as the Mediterranean account for some of the difficulties in the implementation of the CAP, as the largely Northern European measures fail to adequately consider the reality and the particular requirements of the Mediterranean. In the present paper, I will attempt to describe these distinguishing features of the Mediterranean farming systems, to analyze the way Mediterranean products are treated by the CAP, and to trace the possible outcomes for the Mediterranean of the ideas and measures (as far as they pertain to the multi-functional role of agriculture) within the Agenda 2000 framework.

## Particular Features and Development of Farming Systems

The new cultivation systems which allowed Northern Europe to reduce fallow time and to increase yields in the nineteenth century did not gain equal weight in the Mediterranean. The climate did not allow for a rotation of crops in an area with high-yield animal-feed crops. Besides, the modernization for which average-size farmholding was often the vehicle conflicted with the *microfundium* and the *latifundium*, agents of rigidity and inertia for the area. One of the peculiarities of European Mediterranean farming systems has to do with the enduring of an archaic social and agronomic form for large private landholdings, and that for a longer time period than in Northern Europe. Large private landholding diminished in size and importance during the course of the last century, following the agrarian reform of the 1920s (in the case of Greece), the reform after World War II (in the case of Italy) and the post-dictatorship period in 1974 (in the case of Portugal) (Sivignon, 1996).

The extension of irrigated lands, the land reforms of farming plots as well as mass emigration beginning in the 1960s brought on the transformation of rural areas. Irrigated farming, initially confined to riverside basins (essentially irrigation by flooding), relied upon state support. As well, it spread mainly by means of private initiatives, and was directed more by the utilization of water tables. It is now practiced by many small farmers, and it has reached a high level of technical perfection. The modernization of agriculture takes on different forms. Dry hillsides are now planted with olive trees, vineyards, and almond trees. However, it is in the plains where the modernization of agriculture is actually taking place, with an emphasis (climatic conditions permitting) on off-season (citrus fruits and vegetables, which have been irrigated for much longer periods) cultivation intended for export.

In the Mediterranean, the main farming systems exhibit, in general, the following features and development:

*Annual dry cultivations* In dry farming systems the distribution of rain is the most important element. Even in the case of very good rainfall, its irregularity affects crop yields more so than its total amount. Thus, due to the uneven seasonal distribution of rain, there are usually problems in the last stages of growth of winter crops, when, due to the high temperatures of early summer, this growth is often brought to a halt. In the non-irrigated farming systems, the fluctuation of rainfall introduces a major factor of uncertainty for crop yields.

The cultivation of cereals is widespread, but its yield is seriously hampered by the unfavourable soil and climatic conditions. Despite advances in cereal cultivation (mechanization, use of fertilizers and pesticides, new and improved variety strains), the average yield per hectare in these countries continues to differ substantially from that of the countries of Western Europe (for the period 1992-94, the average yield of wheat was 6.410 kg/ha in Germany and 7.500 kg/ha in the United Kingdom, but it only 2.520 kg/ha in Greece and 2.170 kg/ha in Spain).[2]

*Dry tree-cultivations* Trees are an important feature of Mediterranean landscapes. The olive tree, the carob tree, the fig tree, the almond tree, the pistachio tree, the apricot tree, and the vineyards are the most characteristic types and are subject to traditionally extensive management. These trees can endure prolonged summer droughts and irregular rainfall, are able to renew their crowns when ruined, but are sensitive to frost (Quezel and Barbero, 1982). They grow on hillsides and mountainsides, and their cultivation is usually terraced. Terraced landscapes tell of the demands placed on the farmers by the need for land management, these being painstaking and arduous at the least. Today many terraces have been abandoned but are still highly visible, having left indelible marks on many landscapes.

Of course the symbol of the Mediterranean area is the olive tree.[3] Approximately 75 per cent of the world's olive oil production comes from the

Mediterranean countries of the European Union (EU). Today, the trend is toward new plantations, in order to allow the use of machines and the intensification of agricultural practices (treatment of soil, fertilization, pruning, and irrigation), and, thereby, prevent unproductive years.

*Livestock breeding*  The most extensive livestock-breeding systems in Europe are found in the South. They are mainly found in mountainous and semi-mountainous zones, where flocks of sheep, goats, and bovines graze on the wild vegetation of fields, of the *maquis* and of the *garrigues*. Pastoral breeding of sheep constitutes the most common system in these countries. There are also systems for other animals, such as the raising of pigs in the Iberian peninsula, in forest grasslands and sparse tree areas (*dehesas* in Spain, and *montados* in Portugal), and the complex systems of migration of herds and flocks (goats and sheep) between summer and winter pastures (transhumance) in Greece, Spain, and Italy, though these are of limited importance today (Institute for European Environmental Policy, 1994).

As agriculture in Europe becomes more intensive, we can detect significant developments in the breeding of bovines with a focus on purchased feeding methods rather than grazing. The location of modern bovine-breeding grounds is usually determined by the market and/or proximity to big urban centres. Poultry (and egg) production is handled in the same way.

*Irrigated farming*  Mediterranean countries have ages-old experience in water management, as they often have to contend with the scarcity of this resource. In the Mediterranean today, intensive farming means extensive irrigation. Irrigation permits farm units of the Mediterranean countries to meet the demands brought on by the strenuous economic environment by insuring an increase and regularity of yields as well as by allowing a greater variety of crops to be cultivated. The irrigation of land is everywhere more expansive. Thus, in 1993, irrigated lands comprised 19.9 per cent of the cultivated lands in Portugal, 14.4 per cent in Mediterranean France, 22.8 per cent in Italy, and 37.6 per cent in Greece.[4]

In the Mediterranean plains, whether coastal or alluvial zones of the mainland, intensive land irrigation for crops intended for export is practiced Here one already finds an agri-commercial system, for vegetables and citrus fruit (especially early season produce) which is not protected against crises caused by frost, parasites, or the flooding of markets.

By all means the water here used allows for greater production, but it is becoming increasingly scarce. One has to contend, as well, with the inadequacy of water resources, its price, the sur-exploitation of water tables, and the degradation in the quality of water. (Beopoulos and Skuras, 1997).

Summing up, let me note that the difference between the rural structures of Northwestern Europe and of Mediterranean Europe is well known, and that the farming systems of Mediterranean Europe are marked by great variety. Also known is that, on the one hand, there is a coexistence of interspersed settlements, small landholding, multi-cultivation, co-cultivation of agricultural plants and trees, extensive livestock breeding, and, on the other, of aggregates of settlements, large landholding, monoculture of wheat, tree cultivations, and garden produce.

## The Determining Role of the CAP in the Development of Mediterranean Agriculture

The produce coming from the Mediterranean countries of the EU as well as the future of their rural areas is decidedly influenced by the implementation of CAP.

After an initial period of setting out in the 1970s, CAP resulted in significant increase in the productivity of labour and in mechanization as well as in a profound social transformation of the countryside. The structural measures in force were few, but their main goal was to orientate and help modernize 'viable' family farm holdings. The critical rural exodus was absorbed with no difficulty by the other economic sectors. All Community members, including Mediterranean Italy, benefited from these developments. The time period 1970-1980 could be thought of as CAP's consolidation.

It was into this environment that Greece entered the European Community in 1981 (to be followed by Spain and Portugal in 1986). This development increased European heterogeneity and affected the balance of Community agriculture, mainly because it incorporated areas where social and structural problems were particularly acute. Thus the need for convergence between the agriculture of Mediterranean countries and the agriculture of the countries of the North led to actions aimed at the strengthening of this cohesion. Such actions involved programmes such as the Integrated Mediterranean Programmes (IMP), introduced in 1985, for the management of social inequalities and regional differences.

During this time, the decline in active rural population in the Mediterranean countries, both in absolute and relative values of the total active population, was significant. In total, between 1960 and 1990, the active rural population decreased from 42 per cent to 11 per cent in Spain, from 54 per cent to 25 per cent in Greece, from 42 per cent to 17 per cent in Portugal, and from 32 per cent to 9 per cent in Italy. In spite of this, the average active rural population in the four countries of the South (15.5 per cent), still remains much higher than the population of the Northern European countries (5.12 per cent).[5] Of course, the farmers of the North are predominantly farmers (54.3 per cent of the farmers in total are full-time farmers), while the farmers of the South are pluriactive (30 per cent of the farmers in total are full-time farmers) (Lamarche, 1996).

Small farm holdings (the major portion in the South) adapted because of the spread in part-time employment, and the proliferation of non-agricultural sources of income. Of course, in the mountainous and very remote areas, the decline of agriculture was considered inevitable. Meanwhile parallel forms of agricultural intensification emerged near cities and irrigated lands. Here one found comparative advantages, favourable winter temperatures, sunlight, and proximity to markets.

During the 1980s, in sum, the countries of the North and the countries of the South of the EU differ by the manner in which they resorted to Community funds. The farmers of the North, being more productive and more inclined toward heavily subsidized produce (cereals, milk), used a large portion of the credit coming from the European Agricultural Guarantee and Guidance Fund (EAGGF)-guarantee section, while the farmers of the South, facing structural problems (due mainly to the many small landholdings) used a larger portion from the EAGGF-guidance section. This socio-structural support of the South, however, does not in any way make up for the favouritism shown the farmers of the North by the markets (Perraud, 1996).

The spatial dimension of agriculture was incorporated in the EU's regional policy to show its support of more economic and social cohesion by reducing the development gaps (in favour of the less developed areas) so prevalent in disadvantaged rural zones. In this context, a special Cohesion Fund was created, intended for countries with delayed development, among which were three Mediterranean countries (Greece, Spain, Portugal). (Here it should be remembered that the Structural Funds were assigned according to intended objectives. Objective 1 pertains to the regions of EU-15, that is, the total sum of Greece's and Portugal's populations, approximately 60 per cent of Spain's, and over 36 per cent of Italy's. In 1993, the Structural Funds or Cohesion Funds represented 3.3 per cent of Greece's and Portugal's GNP., 1.5 per cent of Spain's, and 1.1 per cent of Italy's (Drain, 1996)).

The CAP reforms in 1992 did not do away with the tendencies in the Policy inclining it in favour of certain sectors of production and certain geographical zones (i.e., towards products and countries of the 'North'). In terms of budget as well as of regulations and arrangement, the CAP involvement was not proportional to the weight of Mediterranean products in the overall production of the EU. Indeed, Mediterranean products constitute a significant portion of the overall EU production. As well, 75 per cent of world production of olive oil comes from the EU, and olive trees cultivation is a concern for roughly two million farm holdings in Southern Europe. Fruit and vegetables make up approximately 16 per cent of Europe' production. Finally, the EU is first worldwide in the wine sector, with 60 per cent of the world production. Furthermore, on a national level, these products are of significant economic and social importance, as they offer direct and indirect employment in zones with serious natural disadvantages.

Exemplary is the fruit and vegetables sector. In 1993, produce from three of the Mediterranean countries, i.e., Italy, Spain, and Greece, made up 57.4 per cent of the European production of fresh vegetables and 73.9 per cent of its fruit production.[6] This sector takes up little land space, employs many farmers, and generates a high added value per hectare, mainly in average-size farm holdings. According to the published statistical data, the EU production of fruit and vegetables uses 4.3 per cent of the farmland (i.e., 5.5 million hectares), but represents over 16 per cent of the total agricultural production, with over 1.800.000 average-size farm holdings. Despite these, the fruit and vegetables sector, in reality, only takes advantage of the guarantee of prices, which are controlled by exchange mechanisms at the borders, and of compensations. Moreover, the concessions made to third world countries for Mediterranean fruit have significant consequences on their price formation. This is unlike the situation in other sectors which take advantage of price guarantees or production subsidies, or both (cereals, beef, etc.). In some cases these sectors are further accompanied by production regulations (land quotas, production quotas, animal-head quotas, etc.).

Even the new CAP reformulations, as they are formulated in Agenda 2000, reveal that, the reform plan (as was the case in 1992) covers only a limited number of agricultural products (European Commission 1998a). The products referred to in the reform (cereals, beef, and milk), represent 39 per cent of the agricultural production of the EU, and 68 per cent of the expenditures of the EAGGF-guarantee section. Mediterranean products are not included in the propositions of the Commission. One ought to note, though, that the CMO of these products, when there is a need for balance, is adjusted accordingly. The 'fruit and vegetables' CMO changed in 1996,[7] the Council had already adopted a common proposition for the 'tobacco' and 'olive oil' CMO after relevant suggestions in June 1998, and the Committee presented a reform proposition of the 'wine' CMO in July 1998.

Of course, the question of why the Commission did not include all products in the reform plan remains. More significant is the fact that the amounts to be allocated towards the reform of the Mediterranean product CMOs, in relation to what is planned for by the CAP reforms are limited. The cost of the CAP reforms is estimated at three billion Euro per year, and that the cost for the Mediterranean products reform is estimated to be an extra 1.2 billion Euro for the time period 2000-2006 (400 million of this intended for wine, and 500 million for olive oil). Indeed, the list of the CAP subsidies for the so-called 'continental' products is expected to be enhanced (Marre, 1998).

## The Need to Provide a New Legitimization for Community Support of Agriculture

Examination of the European Commission propositions of July 1997 and March 1998[8] demonstrates that, overall, the CAP reform seems uncertain. On the one hand, it improves on the 1992 reform through new price reductions which are partly compensated by an increase in direct aid, while, on the other, it attempts to render the CAP more acceptable, since it is intensely disputed, especially by the countries of the North.

This signals a double conflict of a double nature: on the one hand, it supports a specific type of farming relying on intense production which is hazardous for public health (the 'mad cow' crisis alarmed the citizenry); on the other hand, it lends itself to inspection of its allotted amounts and the manner of their distribution in agriculture. Indeed, the weight of agricultural expenditure on the budget of the EU amounts approximately to 50 per cent, with European farmers representing only five per cent of the active population. Moreover, the increased importance of direct payments in the comprising of income from agriculture has led some farmers to elect production of crops stipulated only by these aid packages. Discontent has risen to the surface over this approach.

In the countries of the South, the particularly large numbers of the economically active population working in the primary sector (15.5 per cent in the four countries of the South) biases it toward agriculture. In countries such as Greece, the social acceptance of agriculture is broader because, on the one hand, over 20 per cent of workers, according to the 1991 census, were farmers and, on the other hand, a large part of the urban population are former farmers recently settled in cities, continuing to maintain close ties with their relatives and fellow-villagers, and with farming, since most of them have some kind of land property (which is, though, at times, only symbolic). Since then, the bias has shifted slightly (Beopoulos and Damianakos, 1997).

In the report of general objectives, contained in the document COM (98) 158 final, the European Commission shows that they are taking into account the criticisms concerning some aspects of the CAP, and state that they would like to respond to the actual expectations of society. They emphasize that they are considering a reorientation of the CAP *de novo*, to better reflect the importance of the demands concerning environmental issues, and go on to declare that rural development is becoming the 'Second Pillar' of the CAP, and that their mission is to incorporate agricultural policy to an overall attempt toward the management and arrangement of space, as well as the protection of nature (European Commission 1998a).

These innovative propositions of the European Commission are contained in the projected considerations, and pertain to the creation of an upper limit for demands, the distinctions in the kinds of support, and the integration of rural development in the CAP. Thus, member states will have the ability, within limits, to involve themselves in the manner in which direct payments are made to farmers.

The first distinction is based on the principle of 'eco-conditionality'. Based on Article 3 of the proposition of Regulation 1013/98, member states are invited to determine the particular environmental conditions that are to be respected (depending on the special condition of the agricultural lands in use), and the type of sanctions to be imposed in the case of non-observance. A reduction in direct aid, with the possibility of its being withdrawn (in the case of damage caused to the environment), constitutes one of the harshest types of sanctions (European Commission 1998a ).

The implementation of agri-environmental policy up to the present indicates significant differences between North and South, both in the manner which agri-environmental policy is implemented, and in the way the environment is perceived. Thus, Article 19 of Regulation 797/85, which instituted support for vulnerable zones as regards the protection of the environment, natural resources, as well as the conservation of natural space and landscapes, was immediately implemented only by four countries of North Europe, i.e., England, Germany, The Netherlands, and Denmark (Mainsant, 1992). It was later adopted by other countries, in the form it took in the successive Regulation 797/85, although countries such as Greece and Portugal never implemented it.

Also significant is the implementation of the agri-environmental Regulation 2078/92, which was included in the accompanying measures of the 1992 reform, and constituted an important development towards environmental protection. With this regulation, farmers complying with actions in favour of the environment received support. Based on the evaluation of its implementation by the European Commission at the end of 1997 and in 1998, only 0.5 per cent of Greek, 2.7 per cent of Spanish and 7.1 per cent of Italian farmers participated in these programmes. The corresponding EU-14 average was 13.4 per cent (the data on Germany does not allow for comparisons). Of course, there are significant differences (as in Portugal, where this percentage is reaching 30.6 per cent) but these occur mainly, in the North, where it is very high in the three new member states, i.e., 78.2 per cent in Austria, 77.2 per cent in Finland and 63.7 per cent in Sweden, and in the countries with percentages below the average, such as Belgium (2.8 per cent) and Holland (5.9 per cent). One also finds that, for the EU-14, the landholdings which participate in and belong to regions outside of Objective 1, make up the 20 per cent of landholdings, while, for the regions within Objective 1, this percentage only amounts to a 7 per cent of landholdings (European Commission, 1997b; European Commission, 1998b).

Also to be noted is that environmental problems in agriculture are less acute in the rural areas of the South than they are in the North. In the North, the intensification in farming systems was imposed almost everywhere as the dominant production model. The rapid increase in productivity proved the effectiveness of it, although gradually problems surfaced (economic, social, and environmental) and farmers became more aware of these environmental problems and began looking for alternative solutions with the help of state agents. The case in the South shows a delayed participation by countries in modernization programs for their agriculture and that the kind of development opted for was different from the one in the North (Lamarche, 1996). As a result, in many parts of the South, fewer environmental problems coincide with fewer environmental concerns.

With Article 4 of the regulation proposition, another distinction for support is introduced, one related to employment. Member states will be able to reduce direct support to farm holdings when the available manual labour in use is below a certain limit (translated in work units per year, and determined on a national basis (European Commission 1998a)).

Initially, the increase of production and productivity was accompanied by the elimination of hundreds of thousands of farm holdings in the North. Tolerated as long as farmers could find employment in the city, the elimination of farm holdings is no longer tolerated, as it results in unemployment and the destruction of the rural social fabric. The optional adoption of the employment criterion could increase employment possibilities. In the South, the decline in the numbers of farm holdings is less significant than in the North, and agricultural employment continues. The peculiarity of this agriculture is that it employs a high percentage of the active farming population in numerous small-size farm holdings, many of which specialize in perennial cultivations not found in the North. Of course, the rural population in the South is more aged, and males exceed females in numbers.

The Commission propositions for the rural development sector were presented in the form of a single Regulation (0102/98). These propositions, which limited the objectives of the Cohesion Funds to three, are considered by the Commission to broaden remarkably the scope of possible action, and to alter remarkably the ways of financing. Among the different measures capable of being financed on the basis of the regulation for rural development, are production methods aiming at protecting the environment and safeguarding nature. Indeed, the agri-environmental measures are now to be financed by the EAGGF-guarantee section for rural development.

The Commission proposition to broaden the scope of the CAP and include new concerns, such as environmental protection, (thanks to the mechanism of eco-conditionality, and rural development as the 'Second Pillar' of the CAP), constitutes recognition of the multi-functional character of agriculture – a recognition which, for the first time, is clearly formulated in a EU document. According to the report of objectives, the plan for the CAP reforms has a central objective, which is to give a specific content to what the European agricultural model should be like in future years. The fundamental idea behind a 'European agricultural model' is to allow European agriculture, thanks to the idea of direct support, to continue to concomitantly pursue objectives which are evidently contradictory, i.e., on the one hand, competitive prices, and, on the other, a regionally balanced occupation of the national land and more relevant environmental protection.

In the report of objectives preceding the text of its propositions, the Commission views the quality of agricultural products as the main pursuit of the 'European agricultural model'. In the suggested mechanisms, the issue of quality is placed in the context of enhancement of adaptation and development of rural zones, and it is only viewed from the point of view of marketability.

These propositions bring about new outlets for products and services associated with the management of space, and allow agriculture to fulfill simultaneously its nutritive task and its task regarding the management of space. This can be achieved with the production of quality goods. The concept of 'quality' is steadily gaining ground in agricultural production today through the adoption of practices through a 'productions agreement': use of quality labels, appellation of origin, organic products etc. (Nikolas and Valceschini, 1995). The controlled appellation of origin, as a tool for promoting quality which does not rely on productivity but on the added value of products, is a new way of classifying regions and production locations. Thus the Mediterranean regions of the EU, rich in extensive production systems, product variety, and a reserve of tastes and know-how, could utilize space, production models, and the gastronomic dimension of products, matters which would otherwise, in the name of progress, be disclaimed as archaic.

Agro-tourism is another activity closely associated with rural space. The market for agro-tourism still has room for development in the Mediterranean region, a region with a long tradition and experience in tourism. In most cases this interprets as only offering lodging, while other activities, such as guest reception, board, and recreation, have not been developed yet. One should neither exaggerate potential nor underestimate difficulties. The products and services likely to be offered by rural populations will have to come to terms with the strong competition of the distribution system, the agro-food industry, and the travel agencies. They will only offer real possibilities for economic activity and employment on the condition that local agents will show themselves capable enough to deal with these dimensions, which will not be easy in every case.

Organic farming may be regarded both as an activity which produces quality products, and as a tool showing respect of the environment. Thanks to its nature, it receives double support through Regulations EEC 2092/91 and EEC 2078/92 (which support the transition from conventional farming systems to organic systems). This support allows for the gradual de-marginalization of organic farming and the acceptance of its peculiarities, both by the rural world and by the agro-food sector. The increased demand for organic products in Europe is considered, according to specialists, as a trend of great importance and will lead in the coming years to broadening of the market for these products. Although the presence of areas of extensive farming with a management very similar to the organic, is more evident in the Mediterranean region, it is in the countries of the North (Germany, Austria, Denmark, and The Netherlands), that ambitious programmes of conversion into organic farming are developing – countries with experience and readiness acquired from the application of ecologically oriented programmes in the recent past (e.g., the Opul programme in Austria).

The issue of farmers' compensation for their role as 'gardeners of nature' is often mentioned, and has brought out vivid reactions on both sides, without there being any clear idea on possible ways of initiating such enterprises. Up to present, farmers have provided environmental services free of charge, while, from now on, it is claimed that one ought to consider the cost of these services, and to promote contractual relations between those who express demands for such services, and farmers who provide them. In this category of services belong the preservation of the standard of living, the renewal of resources, and the maintenance of productive capabilities (Gugliemi, 1995).

It is obvious that the nature of activities necessary for the production of quality products is noticeably different from nature of those necessary for the production of high-value rural landscapes. Provision of services by farmers for the production of rural landscapes mainly requires activities influencing their evolution, although many problems and preconditions need to be resolved in advance, in order for this activity to become entrepreneurially practical (Laurent, 1994). Therefore, the question is whether Mediterranean farmers, who have advantageous experience in rural landscapes, will be able to derive use from this new potentially new demand.

The increased sensitivity regarding environmental issues in the countries of the North could be translated as a demand for quality products and for a Nature protected and managed by farmers. Significant, too, is that inhabitants of urban centres in the South are less conscious than their neighbours in the North of the 'new functions' of rural space and, due to their lower income, are likely to express less interest for the new 'agricultural products' in the market (e.g., quality of foods, and recreation services associated with the environment). At the local level, of course, one finds institutions and organizations which could find in these new products and services associated with space, opportunities for complementary resources for agriculture, or a means of utilizing their heritage through the rural development policies.

Many prospects have been created by Agenda 2000 for an agricultural policy which concedes a major role to national and local authorities. These plans interest the South thanks to the possibilities they can offer for the adjustment of policy to the needs of specific areas, and because of their flexibility vis-a-vis the rigid regulations applying to Europe as a whole. There remain the challenges for this kind of development and the risk of a reduction of EU funds intended for the development of less favoured areas (due, in part, to the typic ineffectiveness of public administrations in many areas of the South).

The distinctions of Mediterranean farming systems can be attributed to a natural environment with strong antitheses and a common history. Mediterranean regions generally show the same problems and developmental tendencies, that is, the abandonment of mountainous regions, the utilization and over-exploitation of plains (as far as agriculture and water resources are concerned), and the high concentration of their populations in the coastal regions.

Till now, the CAP has failed to treat all products in the same manner, especially Mediterranean ones. The Commission propositions contained in Agenda 2000 do not change this situation. The disadvantageous treatment of Mediterranean producers is harmful to the cohesion of the CAP.

The CAP is projecting new reforms – granting of direct support on the condition of eco-conditionality and employment, the central importance of rural development and the importance of a more harmonious relationship with the environment. These projections signify the consolidation of the idea of multifunctionality of agriculture which, nevertheless, should be made more concrete within the CAP.

In the context of multifunctionality of agriculture, there are indeed certain outlets being opened for products associated with the Mediterranean region. However, it is not certain that the opportunities offered by means of these new functions and new sources of income for rural areas, can be fully utilized in the Mediterranean region, though the comparative advantages are there.

In general, the issues pertaining to rural space and the role of agriculture are many, and yet, for most of these issues, the new CAP orientations have failed to formulate clear-cut and satisfactory solutions.

## Notes

1   The idea of a farming system used here is based on the definition given by Le Coz: 'A farming system is an organized and targeted set of agrarian structures, techniques of production (of agriculture and of livestock breeding), and exchanges, which develops at a specific production space, in conjunction with the local natural environment, the economy, and the overall economy', (Le Coz, 1990, p. 7).

2   This data comes from the source FAO, 1995.

3   One way to determine the boundaries of the Mediterranean biome is to see in which areas the olive tree, a typical plant of this bio-system, ripens naturally and is cultivated successfully.

4   This data comes from FAO, 1995.

5   This data comes from Eurostat (various years).

6   This data comes from Eurostat (various years).

7   The reform of the CMO resulted in the Regulation 2200/96, which establishes a new Community regime in the sector of fruit and fresh vegetables.

8   In Agenda 2000, which is the reform plan for the CAP presented by the European Commission in July 1997, there is an attempt, through an overall approach, to prepare the Union for its enlargement (European Commission, 1997a). In March 1998, the

Commission announced the sum of propositions of EU actions, which had already been outlined in Agenda 2000. There is a report of objectives preceding the text of the propositions, and an evaluation of the financing following (European Commission, 1998a, document COM [98] 158 final).

## References

Beopoulos, N. and Damianakos, S. (1997), 'Le Cache-Cache Entre la Modernité et la Tradition', in M. Jollivet (ed.), *Vers un Rural Postindustriel. Rural et Environnement dans Huit Pays Européenes,* L'Harmattan, (collection 'Environnement'), Paris, pp. 176-231.

Beopoulos, N. and Skuras, D. (1997), 'Agriculture and the Greek Rural Environment', *Sociologia Ruralis,* Vol. 37, pp. 255-269.

Drain, M. (1996), 'Les Mondes Méditerranéens', in J. Bonnamour (ed.), *Agricultures et Campagnes dans le Monde,* SEDES, Paris, pp. 113-138.

European Commission (1997a), *Agenda 2000 – For a Stronger and Wider Union,* COM (97) 2000, General report.

European Commission (1997b), *Commission's Report to the European Parliament and to the Council on the Application of Council Regulation (CEE) 2078/92,* COM (97) 620 final, 4/12/97.

European Commission (1998a), *Proposals for Council Regulations Concerning the Reform of the Common Agricultural Policy,* Com (98) 0158final/n⁰ E 1052, 18/3/98.

European Commission (1998b), *Evaluation of Agri-environment Programmes, State of Application of Regulation (CCE) 2078/92,* DG VI Commission Working Document, (VI/7655/98).

F.A.O. (1995), *Annual Report: Global State of Food and Agriculture,* Rome.

Gugliemi, M. (1995), 'Vers de Nouvelles Fonctions de L'agriculture dans L' espace?', *Economie Rurale,* 229, pp. 17-21.

Institute for European Environmental Policy (1994), *The Nature of Farming. Low Intensity Farming Systems in Nine European Countries* (eds), Institute for European Environmental Policy, London, WWF – World Wide Fund For Nature, Gland Switzerland and Joint Nature Conservation Committee, Peterborough.

Lamarche, H. (1996), 'Europe du Nord, Europe du Sud le Chassé-Croisé de la Course à L'intensification', in M. Jollivet et N. Eizner (eds), *L'Europe et ses Campagnes,* Presses de Sciences PO, Paris, pp. 77-97.

Laurent, C. (1994), 'L'Agriculture Paysagiste: du Discours aux Réalités', *Natures – Sciences – Sociétés,* Vol. 2, pp. 231-242.

Le Coz, J. (1990), 'Espaces Méditerranéens et Dynamiques Agraires', *Options Méditerranéennes,* Serie B: Etudes et Recherches, no. 2, p. 7.

Mainsant, B. (1992), 'L'Article 19: son Application en France', *Economie Rurale,* 208, p. 136.

Marre, B. (1998), *Rapport D'Information sur le Projet de Réforme de la Politique Agricole Commune,* Assemblée Nationale de France, pp. 159-160.

Nicolas, F. et Valceschini, E. (1995), *Agro-alimantaire: une 'Economie de la Qualité,* INRA – Economica.

Perraud, D. (1996), 'Entre les Contraintes des Marchés et les Hétérogénéites Structurelles des Agricultures Nationales', in M. Jollivet et N. Eizner (eds), *L'Europe et ses Campagnes,* Presses de Sciences PO, Paris pp. 295-316.

Quezel, P. and Barbero, M. (1982), 'Definitions and Characterization of Mediterranean Type Ecosystems', *Ecol. Medit,* Vol. 8, pp. 15-30.

Sivignon, M. (1996), 'Les Systèmes Agraires Européens. Héritages, Mutations, Frontières', in M. Jollivet et N. Eizner *(*eds), *L' Europe et ses Campagnes,* Presses de Sciences PO, Paris pp. 37-53.

Chapter 4

# The Agri-environmental Policy of the European Union – New Chances for Development in the South European Countryside?

Karl Bruckmeier and Teresa Patricio

## The Changing Agricultural Policy of the European Union (EU)

Rural development was traditionally understood as driven by agricultural development in the sense of modernization (Almas, 1998, p. 79) as framed in the Common Agricultural Policy (CAP) of the EU. Since the mid-1980s this policy has changed and been supplemented by the increasingly important structural policies. The Structural Funds represent development-oriented policy instruments with the aim to homogenize levels of economic development in the European regions. In the mid-1990s, the separated policies for rural areas are redesigned under the guiding idea of rural integration or integrated rural development. This paper deals with the integration of rural policies starting from a new component within the CAP, that indicates the change this policy has undergone since the 1980s, that is, the agri-environmental policy as a pillar of the reform of the CAP in 1992 and its continuation under Agenda 2000.[1] The starting point of the analysis is the diagnosis that European policies move from a sectorial approach (agriculture) to one that is more territorial (rural). Further developments along these lines are indicated in the Agenda 2000. In future, the concern of policy will be less to support farmers per se and more to ensure the sustainable production of environmental and other public goods together with the prosperity of the wider rural population. Support is therefore likely to become increasingly focused over time on environmental measures and on rural development policies to develop the capacity of rural areas to support themselves (Shucksmith and Chapman, 1998, p. 225).

Research has played an increasingly important role in the development of the CAP (Dent and McGregor, 1995); it was mainly influenced by economists who saw agricultural modernization in terms of a productivist model with intensified resource use (Ward, 1993). With the reform of the CAP in 1992 this model came to be seen more critically. Two trends have become dominant: a price reform, signalling the opening of agricultural production to market forces and a series of measures to support environmentally sound agricultural production which form the basis of agri-environmental policy. This policy is linked with other goals of the policy reform such as reduction of agricultural surpluses in a 'magic circle' of mutually reinforcing environmental, market and income policies. The guiding idea is to keep farmers on the land for reasons of nature and landscape conservation (Scheele, 1996, p. 4).

Environmental measures as formulated in Regulation 2078/92 of the EU have not been without precedents, as there have been prior programmes dating back to the mid-1980s (Baldock and Lowe, 1996). The regulation provides financial support for farming practices to reduce the polluting effects of agriculture as well as to promote environmentally friendly extension of crop, sheep and cattle farming. It further encouraged environmentally friendly methods of agricultural land use, upkeep of abandoned farmland, long-term setting-aside of agricultural land, land management for public access and leisure activities, education and training of farmers for environmentally friendly types of farming. These agri-environmental measures represent a combination of new instruments in agricultural policy; individual management contracts between farmers and state, individual and voluntary participation of the farmer, and compensation payments for income losses through reduced yields. This combination indicates a change of policy instruments away from 'hard', regulatory and government based instruments (like permits and bans) to 'soft' incentives, persuasion, or negotiation-based policy where the individual farmer, his orientations and decisions are of increasing importance. Subjective factors, for example, attitudes, orientations, values, objectives of the agricultural producers, factors which vary greatly according to regional and cultural traditions, become important in either supporting or blocking the implementation of the agri-environmental policy.

As far as the national and regional adaptation of agri-environmental policy is concerned, Regulation 2078/92 conceded to the member countries a large degree of freedom in the formulation of national, regional and local measures. Furthermore, the programmes in the member countries indicated different views about environmental protection in the agricultural sector (Whitby, 1996). The implementation regulation of April 1996 was a reaction against the splintered and incoherent implementation of agri-environmental programmes in the first phase of the reform. The new regulation attempted to limit the variation in national and regional programmes by claiming more precise criteria for extensive agriculture, good environmental practice and calculation of compensation payments. Although such standardization may be justified under criteria of equity and efficiency, it illustrates the contradictory aims in the policy of the EU. Agri-environmental policy opens paths to more regionalized, less centrally controlled development and when this process has gained momentum, it reacts by standardizing the criteria for 'good' rural development in the traditional way of powerful top-down intervention.

Not long after the start of the policy reform in 1992, new scientific and political debates followed, posing the question of whether to reform the reform, some proposing 'market radical' solutions, others proposing a strengthening of integrated policy approaches for rural areas beyond sectored ones (Buckwell, 1996). What the Commission has taken from these discussions and brought into its own report for

Agenda 2000 (European Commission, 1998) was less consequent, but it indicated the further reduction of agricultural producer prices and it evoked the question of how long a policy of compensation payments can continue.

## Agri-environmental Policy and the Transition to Sustainable Development

With the reform in 1992, the criteria for agricultural modernization changed. This happened in a situation of a differentiation of farming types that offer different solutions to the management of environmental problems. Modernized, intensive farming is the mainstream model, but agri-environmental policy with the core of Regulation 2078/92 is designed to support four types of environmentally sound agriculture.

*Traditional forms of agriculture* include low external input agriculture as well adapted to local ecosystems (hitherto neglected forms of agriculture under the paradigm of economic and technical modernization).

*Extensive agriculture* is similar to traditional agriculture with regard to natural resource use, but vaguely defined in terms of its ecological accommodation to local and regional conditions of production ('extensive agriculture' is a category of the agronomic sciences, not one from the practice of farming).

*Integrated farming* is a model derived from integrated pest management; it is often – but inadequately – seen as a compromise for the majority of farmers to produce environmentally sound products without converting to organic farming.

*Organic farming* is a consistent approach to farming following ecological criteria of ecosystem management (which has existed hitherto mainly as 'ecological movement farming' without support through agricultural policy).

All four variants of sustainable agriculture include crop and animal production.[2] However, all these options require specific social and economic conditions in order to become viable. They cannot be created as political utopias in a social vacuum. The success of these production systems requires regional and local markets and adequate trade systems for agricultural products; it requires, furthermore, ecologically aware consumers who are able and willing to pay for products from organic farming, regional food production industries in economical and socially sound rural communities, agricultural landscapes which represent ecologically well-managed ecosystems and the existence of societal consensus about agriculture and the use of natural resources in rural areas. So far, the support of traditional agriculture and animal husbandry is an experiment – more of political aspiration and less of widespread social practice.

   There is no coherent approach to agricultural and rural development underlying the reform of the CAP, although the principle of sustainable development has become the overarching framework of European policies in the 1990s. The idea of sustainable development of rural areas is formulated in some scientific documents (SRU 1996 for Germany) and, as a global development strategy, in Agenda 21 where its agricultural variants included organic farming, integrated farming and extensive farming as systems of environmentally sound agriculture. Several variants for a 'sustainable transition' (European Commission, 1995; Brand, 1997; O' Riordan and Voisey, 1998) have been discussed, with two principal alternatives as background models: sustainable development as a policy-driven approach and as a culture-driven approach.

The first approach describes the continuity of 'policy as usual' in a state-controlled, interventionist model with the following deficient premises: an overestimation of the capacity of governmental and political institutions to influence development; the standard diagnosis of the relative economic and structural backwardness of southern member countries of the EU; a one-dimensional development concept of early/late modernization; the construction of standardized rural development policies on the basis of macro-economic and structural data which give little room for different regional development paths; the traditional view on ecosystem degradation through agriculture which tends to be combined with a  technocratic view of denying the existence of negative externalities or following the ideas of compensation and repair.

This approach is informed by structural data and development is conceived at the aggregate levels of macro-economic data and socio-economic structures. A thinking of sustainable development in terms of some overall visions of a sustainable society (Moffatt, 1996, p. 164ff) is underlying this structural model. And there is an inclination to standardize sustainable development into one or more models that seem to be coherent from a policy implementation view. Furthermore, this policy-driven approach is based on the implicit – however unclear – assumption that sustainability can be achieved by the ordinary means of politics and power. Socio-cultural traditions, local knowledge and social factors do not count much here.

The second, or culture-driven approach, supports ideas such as the following: development is no longer seen as a replication of the standard path of Western industry-based development but as a social process which allows for subjective capacities to unfold (micro-policy) and a plurality of development paths to follow (economic, social and cultural differentiation); development processes should be broken down from the level of societal systems to micro-levels of socio-cultural conditions of development as specified in terms of local socio-cultural systems, actors, actor-networks, social movements, interest groups, structures of civil society; different geographical and physical conditions of agriculture are taken into consideration as well as socio-economic conditions under which agriculture is done; regional development patterns and regional integration of development are combined with this approach (supporting the adaptation of agriculture according to local ecosystems) and the global change processes are seen critically from the point of view of sustainable management of local rural ecosystems.[3]

This approach allows for a better consideration of regional contexts, social actors, and their interests and capacities, as, for example, in the variants of endogenous development, local development and 'grassroots' development. In the 1980s such thinking emerged in political discourses about rural development in European countries as well as in international development discourses (Cernea, 1991). In terms of politically conceived development strategies the ideas are often merged with simplified forms of 'bottom-up approaches' and participatory development that entered policy programmes of the EU, for example, with the Community initiative LEADER (Ray, 1997). We follow a modified interpretation of the culture-driven approach, which does not reduce it to utopian ideas of local development. Furthermore, critical normative ideas about civil society and social movements come into discussion here for the purpose of mobilizing local knowledge and traditions against the growing influence of global systems (or against the 'colonialization of the human lifeworld through imperatives of systems' in the words of Habermas). The view of rural development as micro-level and culture-driven processes should be specified in regional approaches that take into account power structures, economic influence, and conflicts in development processes. Regional and local approaches cannot be de-coupled from the large-scale political, economic and social processes.

When the first model is identified as one describing the orientations of non-local actors (states and economic corporations) under conditions of economic globalization, and the second one as that of describing local resistance to destruction of local social and ecological systems, we can find several connections between the approaches. The critical question for both approaches is that of the coherence between processes at different levels of development. From an ecological perspective, the variants of interaction can be classified under the principle of coherence between macro-level policy, micro-level social processes and the evolution of ecosystems. Using a more familiar expression from ecological economics, such coherence can be called 'the fit between ecosystems and social systems' (Folke et al., 1997):

1. *'No fit'*: Coherence and interaction between the macro-systems of policy and economy, local social systems and ecological systems are lacking and 'ecological marginalization' (Kousis, 1998) may result, that is, the imperatives of macrosystems (the globalizing economies, national and international political systems) dictate the roles that rural regions play as backyards of industrial development centres with the consequences of progressing degradation of local ecosystems.

2. *'Bad fit'*: Insufficient coherence between the three system levels is given in cases when ecological marginalization is temporarily hindered or reduced by policy programmes and compensation payments but does not result in a take-off of sustainable rural development. This happens with the agri-environmental policy in the majority of rural areas in Europe; local, autonomous development capacities are not set free and degradation of local ecosystems is not consequently prevented.

3. '*Good fit*': Coherence between the three system levels is obtained in cases when ecological and social destruction of rural areas is countered by a successful strengthening of traditional forms of agriculture. This can happen as an impact of agri-environmental policy where it is implemented in regions with traditional extensive agricultural systems to support such agricultural systems which are still widespread throughout Europe (Bignal and McCracken, 1996). In this case the continuity of traditional forms of managing ecosystems through agriculture helps to prevent the degradation of local ecosystems. Unfortunately, the map of extensive agricultural systems in Europe is not identical with that of successful implementation of agri-environmental policy.

4. '*Optimal fit*': Lasting coherence between the three systems levels is given, for example, when successful implementation of ecologically accommodated agriculture becomes a mainstream approach and the extremes of segregation and marginalization of rural areas can be stopped as well as a progressing degradation of ecosystems. This positive variant is one where agri-environmental policy must be supplemented by approaches and factors other than agricultural development; agriculture must be embedded in regional strategies of development that can be based on pluriactivity and diversification.

A development concept supported by the agri-environmental policy is that of regionally integrated development in the North and the South where the regional economic disparities should be reduced – not through increased transfers but through the stimulation of local and regional development capacities. This would imply a shift from the policy-driven approach to the culture-driven approach of development. At present, such socially anchored, region-based, balanced and integrated rural development has few chances – more prevalent are conflicting and contradictory policy objectives. In Southern Europe, namely in Spain, government as well as agricultural unions follow a double goal of completing modernization and reducing productivity gaps to reach the economic standards of the North simultaneously with the promotion of sustainable rural development (Moyano, 1995, p. 362). Such fragmented policies may contribute to regional segregation but not to regionally differentiated sustainable development. Broader scenarios that include the development of southern member countries of the EU by strengthening inter-regional integration (Gaudemar, 1995; Waniek, 1995) do not provide convincing arguments for regional development. Such scenarios are constructed from macro-political and macro-economic data with little sensitivity for regional specificity.

## Agri-environmental Measures in Four Countries

### The Implementation Process

The following descriptions of the introduction and initial impact of the agri-environmental measures of the CAP-reform refer to four countries: France,

Germany, Portugal, and Spain. Main sources of information are some recently finished analyses of the agri-environmental policy and sociological studies of rurality and environment (Billaud et al., 1996; Jollivet and Eizner, 1996; Whitby, 1996; Jollivet, 1997). Information on the structural differentiation of European agriculture is taken from Limouzin (1996). Supplementary information sources are used for single countries or regions (Moyano, 1995; Hoggart, 1997; Izcara Palacios, 1998; Alphandery and Pinton 1998; Kousis, 1998; Rodrigo, 1998).

## 1. France

In France the introduction of Regulation 2078/92 has contributed to a rather quick acceptance of the agri-environmental policy by the main agricultural unions and interest groups. The agricultural community accepted a policy which is built on the premise that there are serious environmental problems caused through modern agriculture. This argument was denied by most parts of the agricultural community until the early 1990s.

Local measures are of some significance in France, but these are of a heterogeneous type and cannot easily be compared with that in other countries. The environmental measures most in demand by farmers are those which require a reduction in agricultural inputs – this is similar to the participation structure in Germany. Great interest among the farmers was found, especially for the grassland extension programmes. That is, many farmers chose to participate in measures that necessitate only minimal changes to their production systems.

The reasons why farmers participate in the measures are very varied and linked to their form and degree of production specialization. However, one motive, which dominates in every country discussed in this paper is that of economic gain – additional support and payment for agricultural production are reasons to apply the measures – although the amount of payments may be limited and a significant effect on income is achieved only in combination with other agricultural support programmes (that may not be coherent with the agri-environmental programmes). Ecological motives – environmental concern, nature or landscape protection – seem to be of secondary importance.

The correlation between different types of specialized production systems, their participation in measures and their motives for participation requires detailed empirical analysis.[4] Beyond predominantly economic motives for practicing agri-environmental measures combined with an increasing general environmental awareness on the part of French farmers, there appears to be no generalized or coherent motive as to why farmers participate in agri-environmental programmes. Increasing environmental awareness amongst farmers can be seen as a turning point in the agricultural community's long dominance in France. Another turning point may be that the farmers have begun to realize that the agricultural profession must negotiate its production methods with other groups in society.

## 2. Germany

Since 1990, with the integration of East German agriculture, a split has developed in German agriculture. Three structurally and regionally distinct forms of agriculture can be found in Eastern, Northern and Southern Germany, with those in Northern and Eastern Germany retaining the competitive advantage:

- large-scale agriculture in Eastern Germany, developing forms of entrepreneurial and commercial organization which are no longer based on family farming;
- modern, specialized and intensive family farming systems in Northern Germany;
- small-scale, traditional or partially modernized family farms in Southern Germany, often run on a part-time basis, and continuing to face poor long-term economic success. The exception is where they are able to find temporary market niches for new products or specialist crop production (wine, tobacco, etc.), which, for climatic reasons can only be grown in limited areas.

Unlike in France, a main effect of the measures was not a change of attitudes in the farming community, as this had gradually happened since the early agri-environmental programmes in the 1980s. The statistics indicate that the measures have been important in terms of the area to which they have been applied and the number of farmers participating. The implementation differs between the regions. Every region (*Bundesland*) has a programme of its own, its own conditions of participation and levels of payments.

Farmers' interest and motivation for agri-environmental measures varies considerably. Instrumental-economic motives (receiving additional income from one of many support programmes in which farmers can participate simultaneously) do prevail, whereas ecological motives appear to be less important. A lack of interest of farmers in the programmes can be observed in Northern Germany where modern, intensive and specialized agriculture is concentrated. Most success in terms of farms participating in the schemes has been observed in the Southern German regions where small-scale family farming is still widespread. In some regions in East Germany the large areas covered by the programmes are a direct consequence of the participation of large-scale holdings resulting from the reorganization of socialist agriculture. Such holdings have large reserves of agricultural land and can expect large amounts of additional income as a result of applying the measures to areas covering hundreds of hectares and more.

The ecological impact of many measures is limited. In some regions and for certain types of farms which already produce extensively, participation of farmers may lead to a 'free-rider effect' (where the participation does not generate additional ecological benefits). The overall area and number of farms covered by the regional agri-environmental programmes is still low in comparison to the total agricultural area (c. 20 per cent of grassland and 5 per cent of arable land).

## 3. Portugal

The general development trend in Portugal is towards the decreasing economic importance of agriculture. In large sections of the country, mainly in the interior, agriculture has not modernized to produce toward the market. Traditional agriculture survived through a combination of family strategies such as alternative forms of family employment which permit subsistence family farming. Portuguese agriculture, similar to other Southern European countries, is largely characterized by a structure of family farming that relies on outside income to maintain the farm.

The agri-environmental problems tend in Portugal to either be denied or minimized as local and specific. Regulation 2078/92 was the first measure which made agri-environmental issues the subject of a national policy and challenged the general impression that such problems were local. The Portuguese 'translation' of Regulation 2078/92 proposed a large number of measures dealing with agri-environmental issues within a rural development perspective. The focus of the regulation, and the largest number of measures, dealt with human and physical desertification that was identified as the main agri-environmental problem in the country.

Given the various patterns of development, regional differentiation and environmental impact, a redefinition of the measures of Regulation 2078/92 is required to make them functional in terms of agricultural development. In this way the measures will provide additional means of social and economic support to agriculture aimed at reducing rural exodus or 'social desertification' of the countryside. Furthermore, preservation of traditional forms of agriculture has more significance within the implementation of the regulation. The focus was not reconversion to extension but the maintenance of existing traditional, extensive forms of production. While intensive agriculture was included in the package of agri-environmental measures, it was less discernible, introduced later, and attributed a lesser budget.

Regarding the results of the implementation of measures in Portugal, there is a stronger participation from regionally specific forms of traditional agricultural production than can be seen in Spain. The measures are, however, implemented in a formal top-down approach and farmers react rather passively, in a 'wait-and-see' attitude. The main measures have targeted traditional agricultural systems of production in a rather diffuse manner, which reveals the measures to be essentially a subsidy for the preservation of land use or as an additional contribution to the existing state pensions.

## 4. Spain

As in other South European countries, agriculture in Spain is of greater economic importance than in the northern member countries of the EU, although the agricultural population has decreased. Spanish agriculture, due to the physical variation of the country, is regionally specialized. Cattle rearing has developed in the north and west, extensive farming with dry-land crops has evolved in central regions, especially Castile, and intensive agriculture (fruits and vegetables) is concentrated along the Mediterranean coast (Pérez Yruela, 1995, p. 277ff; Limouzin, 1996, p. 59ff).

The measures were implemented rather late, as in Portugal, and even in 1996 not all Spanish regions were included. This is partly a consequence of the specific political structure (regions with a high degree of autonomy in a nevertheless centralized state), and partly a consequence of the high regional diversity in terms of climate, availability of water and soil quality. The following points give a summary of how the agri-environmental measures are implemented in Spain:

1. An important trend – not counter-acted by the agri-environmental measures – is that of continuing modernization aimed at producing intensive production oriented agriculture.
2. Farmer participation is low for all types of measures.
3. The programmes, horizontal and vertical measures included, do not tackle the main environmental problems from which Spanish agriculture is suffering, i.e., erosion and dry soils. Water is the critical factor of Spanish agriculture and not soil. The special problems of irrigated areas (tripling in size since 1950, with 3.4 million hectares of irrigated land which use 80 per cent of available water resources) are not touched upon by the agri-environmental measures.
4. The budget for the measures is too small for them to become effective instruments supporting social and economic development in rural areas.
5. The measures are not integrated with other policies targeted at rural areas and do not therefore form part of a regionally integrated and coherent policy of rural development.

In Spain the implementation of the agri-environmental measures will widen not only the gap between intensive and extensive agriculture. In addition, the geographical diversity of agricultural systems produces a great diversity of environmental problems. These problems are not only the result of modernization and intensification, but sometimes also the result of non-modernized farming with environmentally damaging techniques (e.g., wasting irrigation water) or in certain areas land abandonment (Garrido and Moyano, 1996, p. 93). Spanish agriculture, in comparison to Portuguese, is more physically, politically and socio-economically differentiated, which may be one of the reasons why it has proved more difficult to adapt and implement the agri-environmental measures.

*Comparing Agri-environmental Measures in Four Countries*

The short descriptions given above show different ways of adapting the measures to ongoing rural development processes in the four countries. The four cases can be taken as illustrating the dominance of socio-cultural effects (France) or of economic effects (Germany), of a socio-economic policy of stabilizing traditional agriculture (Portugal), and an example of an institutionalized contradiction between ongoing mainstream modernization and ecological adaptation (Spain).

*Dominance of socio-cultural effects – France* In France, agri-environmental measures have, to date, had little influence on changes in agricultural systems. It is therefore impossible to clearly assess their long-term impact or their wider economic, social and ecological implications. However, it can be said after the first few years of implementation that, in socio-cultural and political terms, the impact of the measures has been both to change the identity of and legitimize agriculture (Billaud et al., 1996; Alphandery and Pinton, 1998). For these reasons the measures have found great social resonance in France. This is partly explained by the coincidence of the introduction of Regulation 2078/92 and its regionally and locally profiled measures with an ongoing political debate on decentralization. Furthermore, the negotiation and co-operation involved in implementing the measures stimulates innovation in agricultural policy and decision-making at both regional and local levels.

The significant consequences of agri-environmental policy in France have been those in the on-going social and political debate about agriculture. This is in contrast to Germany, where the measures are discussed more in terms of their economic and ecological impacts, for example, how they influence existing agricultural systems and how far they contribute to the emergence of sustainable regional agriculture. Whereas in France the agri-environmental policy is contributing to legitimizing and creating a new identity for agriculture, in Germany it is helping to de-legitimize mainstream agriculture. Although this de-legitimization may support the change towards environmentally sound agriculture, its dominant blocking effect is to contribute to a crisis of identity facing farmers, whose income no longer results primarily from their productive activity. Co-operative implementation in France is reducing the dominant role of agriculture in managing rural space and landscape at local levels. However, the policy still lies in the hands of the agricultural community. The environmental policy networks, nature protection groups or new environmental movements do not play a significant role (as in Germany) in the negotiation process at regional levels.

*Dominance of economic effects – Germany* After the first years of implementation of the agri-environmental measures in Germany, both the intended and the unintended effects of the reform of the CAP are to some extent clear: The intended impact, i.e., that farmers change their methods of production, has been limited, as has the ecological impact of improved landscape, soil and water quality. However, the variation of regional programmes makes an exact quantitative account of the effects impossible.

Among the unintended and counter-productive social impacts of the reform is the crisis of legitimacy in the farming community. Many farmers perceive the policy reform to be a step-by-step reduction of political support for domestic agriculture. Changing transfer mechanisms and compensation payments challenge the traditional role and legitimacy of agriculture as food producer. For farmers, a critical point is reached when transfer payments become separated from the productive performance of the farm, that is, when the economic justification of farmers as food producers disappears. Farmers, influenced by the CAP to modernize farming and to build an entrepreneurial identity of food producers still find it hard to accept that they should be paid not only for their economic performance but also for their ecological performance in terms of conservation of landscape and natural resources.[5]

Another of the unintended economic effects of the reform of the CAP is the different 'free-rider effects' that agri-environmental measures have had. In East Germany large production units have gained large amounts of additional income as a result of the large areas they have been able to enter into the programmes. Similar effects have been observed in Southern Germany where only minor changes in agricultural production have meant that farmers derived additional income without producing significant additional environmental benefits. In Northern Germany farmers' interest in the measures is limited.[6]

A third unintended economic effect of the reform has been the increasing importance of set-aside measures. This leads to a scarcity of productive land in agricultural areas, causing stagnant or increasing prices for leasehold and increased transfer of rents from producing farmers to non-producing landowners.

*Socio-economic stabilization of traditional agriculture – Portugal*  In Portugal the survival of traditional farming is discussed when it comes to the agri-environmental policy, as this policy may become a means to create socio-economic conditions in which traditional agriculture can continue. Traditional farms can survive when they are able to integrate farming activities with those of rural development and the social functions of rural communities. Implicitly this is a trial to keep these farmers out of the mainstream modernization trap. Some of the other activities, associated with this multi-functional role are the production of regional or high quality products, agro-tourism, leisure activities, craft and small-scale business enterprises as well as natural parks and habitat reserves. How far such alternative activities are undertaken by farmers and contribute to economic and social stabilization of rural areas is not yet sufficiently clear, although rural development programmes do seem to contribute to new income-producing activities for farmers.

In assessing the political, social and economic chances for new local and regional development models, the following points are important:

- The active role of the Portuguese Ministry of Agriculture in developing agri-environmental measures as a means for slowing down social and physical desertification of rural areas has been a decisive component of their success, although still not sufficient to make them efficient instruments of rural development.
- Case studies in some regions of Portugal and other South European countries show that the trend towards marginalization of agriculture can be halted or slowed down, not just by European or state policies, but also by the active resistance of rural populations (Bazin and Roux 1995, p. 345).
- The observation that the pluriactivity of the rural population in Portugal has reduced the dependence on agriculture as the economic basis of rural areas

(Baptista, 1995) underlines the role that non-agricultural activities could play in rural development in the future.

All these observations seem to indicate that Portugal is a special case of agricultural development, however, within the limits of the South European structures of regionally differentiated and predominantly small-scale farming. Some of the unintended impacts of the policy reform, such as unequally distributed advantages for certain groups of farmers, have also been reported from Portugal.[7]

*An institutionalized contradiction – Spain* Spanish agriculture is more advanced on the path of modernization of production systems than Portuguese agriculture. Agri-environmental measures are introduced into a regionally and economically more diversified agriculture and there is no such coherence in goals as in Portugal.

> In the agri-environmental field, the Spanish Government's actions have followed the steps indicated by EU and generally Spanish farmers have not shown much interest in them. Insufficient information, the lack of participation of farmers' unions in their implementation and the small amount of money devoted to these aids can be some of the reasons for this lack of interest. However, the new environmental concern introduced with the Common Agricultural Policy and more specifically the accompanying measures in the CAP reform of 1992, have served to energize the debate within the agricultural sector, specifically within the farmers' unions. This debate embraces the relationships between agriculture and environment and it has taken the notion of sustainability as a reference framework to put order on it (Garrido and Moyano, 1996, p. 104).

The effects of the policy reform and of the agri-environmental measures vary too greatly between regions to allow for any generalized country-by-country comparisons. However, some of the impacts that do not depend upon regional specificity can be identified already:

- The measures did not break the dominance of intensive agriculture.
- For different reasons participation of farmers in agri-environmental measures is limited in the majority of countries. The situation in Portugal where most farmers participate in the measures is an exceptional case.
- The ecological impact of the measures, in terms of achieving environmentally sustainable agricultural systems, has been limited in all four countries.

By comparing how the measures are implemented in four countries, the following factors can be singled-out as important in the development of regionally specific agricultural systems:

1. Support of traditional agriculture and rare breeds may become more important for Southern Europe, where agro-biodiversity is an important question.
2. Integrated rural development (encompassing social, cultural and economic development) is the common denominator for northern and southern variants of the agri-environmental policy.

3. Organic farming as a regional nucleus is an ideal of ecologically adapted agriculture which has rapidly grown in the past years but which remains at a low level.
4. The rise of new approaches to the development of the countryside indicates a non-agricultural component of agri-environmental policy; here the agricultural landscape is increasingly becoming a resource for the non-agricultural population and their consumption of nature in recreational activities; the social capacity and participation of the local population in landscape development needs often to be strengthened (Volker, 1997).

## Deficits and Limits of the Agri-environmental Policy

In terms of promoting sustainable agriculture and integrating social, economic and ecological criteria under the broader framework of sustainable rural development, not all agri-environmental measures are efficient (SPU, 1996). In addition, the measures seem to be targeted primarily at the type of agriculture that has developed in the northern member countries as intensive modernized farming producing large surpluses. This provided for the main justification of an agri-environmental policy. The archetype of this diagnosis can be found in the early documents from 1985 where the European Community argued for a new development of agriculture (Europäische Kommission 1985, p. 50ff) under the increasing pressure of environmental damage resulting from agricultural practices. The argumentation was rather vague with regard to the southern member countries: they were included in the diagnosis of increasing environmental damages through agriculture although it was seen that the negative environmental impacts of intensive agriculture and agro-chemicals did not cause the main environmental problems here.[8]

The support of traditional crop production or animal husbandry (local breeds) is of little importance in Germany and France, but more significant in South European countries. However, the question is whether the existing traditional systems can become the regional centres and moving forces for agricultural development. The strengthening of traditional low-external input farming systems (WWF, 1994; Bignal and McCracken, 1996) has not been due to the effects of agri-environmental measures, but a synergetic effect of different endogenous factors as support from local actors and groups, ecological movements and non-governmental organizations. Organic farming will remain a minority path in all three countries and generally of less importance for Southern Europe; the agri-environmental policy does not change this situation. Support for integrated farming which represent the de-intensification path of rural development (not necessarily a component of integrated rural development) is a model implicit in some regional programmes in Germany but is of little importance for the South European countries. Given all this, it can be concluded that the agri-environmental measures have not opened new or additional paths for ecological development of agriculture in Europe, but have only reinforced the already existing alternatives.

From the observation of the implementation of the agri-environmental policy in four states result the following conclusions:

1. Agri-environmental measures do not represent an optimal 'policy mix' but more a box of multi-purpose tools with too many different and ambitious objectives. If one objective is to encourage animal welfare and extensive production, the measures have not yet had a strong influence on mainstream agricultural modernization that continues to develop genetic engineering as a way to further intensification.
2. It is doubtful whether the agri-environmental measures can have a significant impact on intensive and highly specialized agriculture or livestock farming in terms of reconverting it into more environmentally sound agriculture.
3. Support for traditional forms of agriculture would give existing regional production systems a better chance. However, it remains to be seen how traditional agriculture can become the driving force for sustainable agricultural development.
4. The measures can become important means for landscape development and the conservation of ecologically accommodated agricultural landscapes under specific conditions only, when they are integrated with further conservation programmes.

Regionally segregated development may occur – as whilst it may not be strengthened by the reform of the CAP it is not prevented by the implementation of agri-environmental measures. For other forms of production the effects may be limited too, as has been said above with regard to organic farming. The support of traditional forms of agriculture will give some local and regional production systems chances for longer survival than under prior policy. If the agri-environmental policy is to become a means for supporting traditional and extensive farming, it needs to be more than a programme for subsidizing the use of certain production factors. A new policy is required to support the social and economic structures of agriculture that have to be protected against the competitive power of modern mainstream agriculture. Such a policy implies a change of the criteria for agricultural modernization and rural development. Never before in the history of the CAP such a far-reaching change has been thought of. The process should be seen in the long run with outcomes that represent examples of regionalization in the positive and the negative sense.

## Agri-environmental Policy in the Framework of Agenda 2000

With the implementation of the agri-environmental measures expectations have been created that this policy will continue in the long run. The idea of integrated rural development has evolved as a trans-sectorial framework for agricultural policy. With such a policy the links between landscape development, rural society and diversified agricultural production need to be developed. A new rural identity

(Jollivet, 1997) may evolve when the importance of multi-functional activities within rural society is recognized. In contrast to this rather optimistic view is a more sceptical view of the chances for future development that is based on evident structural and economic constraints and the limited interest of many farmers in new forms of agriculture. Such scepticism is not removed with the Agenda 2000, although the agri-environmental measures continue to be a main innovative component of EU-agricultural policy.

Through Agenda 2000 the framework for the implementation of the agri-environmental measures has been broadened towards integrated rural development as it has been declared in an act of symbolic consensus between governmental and non-governmental actors in Cork 1996. A multi-sectorial and integrated approach counts among the principles of the Agenda (Tronquart, 2001, p. 10), but it appears less clearly in the Council Regulation on Rural Development (1257/99) where the measures for the policy of rural development are specified. In this regulation the integrated approach is reduced to a compromise between the further dominant agricultural and agri-environmental measures and the added measures for the general development of rural areas under article 33.

The recent comparative analyses of agri-environmental policy (Buller et al., 2000) and of the new rural development policy under Regulation 1257/99 (WWF 2001) do not indicate that the problems discussed above for four countries in Northern and Southern Europe have vanished. Only the changing context of the agri-envrionmental measures within the framework of Agenda 2000 becomes clearer: from a minor component of agricultural policy the agri-environmental measures have become a major component of a policy aiming at integrated rural development. Such a policy is not yet dominant practice, it is still hindered by the dominant interests of conventional agriculture.

## The Future of Agri-environmental Policy

Although the promises of integrated rural development are not yet realized, in northern and southern member countries of the EU, rural development will be more and more influenced by non-agricultural groups, actors and factors. The successful transitions to sustainable agriculture in which these groups may play a key role seem to be less national and political, more regional and socio-cultural ones. The unsolved problems, knowledge deficits, and barriers to be overcome before agri-environmental policy can become an effective tool for integrated or sustainable agricultural development can be summarized as follows.

*Unsolved scientific questions* come with the evaluation and assessment of long-term impacts of the agri-environmental measures (Buller et al., 2000). Evaluation should be multi-dimensional, referring to the social, economic, political and ecological differentiation of agricultural production systems. The choice of suitable frames of reference (individual farm, similar production systems, ecologically homogeneous areas, political and economic regions and countries) is difficult and only insufficient data from several levels are available. It will be important to evaluate quantifiable economic and ecological effects but also qualitative socio-cultural effects on orientations and attitudes of farmers. Evaluation has so far tended to neglect such social and subjective components, and intensive contact to the target groups was not sought. The new guidelines for the evaluation of rural development in the years 2000-2006 (European Commission, Directorate General for Agriculture 1999) demonstrate that the evaluation approach tries to model the complexity of the rural development processes by being ever more complex – the key evaluation aspects include effectiveness, efficiency, relevance, utility, sustainability and coherence (European Commission, Directorate General for Agriculture 1999, p. 16). However, the approach is still not more in contact with the subjects and target groups of rural development.

*The role of socio-cultural factors* is of increasing importance for the implementation of the agri-environmental policy, as is the support from the general public, the farming community and the integration of farming in local markets. In regions where extensive, traditional or organic farming receives social support through integration into local food markets this was the case before the reform of agricultural policy. The situation has changed due to a combination of factors, of (a) internal influences on agricultural policy (especially consumers' wants and needs, health aspects of food production and consumption) and (b) external influences (environmental aspects, rural tourism, regional and integrated rural development). In such regions the chances for further successful development seem better than in regions where changes happen recently and abruptly as a reaction to changing policies of the EU. To determine the influence of socio-cultural factors on rural development, specific research is required which addresses the social subjects of rural development, the different orientations, visions and values of resource user groups in rural areas; simple opinion polls, revealed preferences in the form of 'willingness to pay'-analyses, and observation of factual economic behaviour will not suffice for this purpose.

*Policy approaches* of importance are efforts to develop agriculturally and environmentally integrated, supra-sectorial policies for rural areas to integrate agricultural and environmental policies that come into practice with the agri-environmental policy. How such policies can be built up and implemented can be learnt from regional experiences in coastal areas. There is a growing and worldwide trend to integrated policies for coastal zone management. Strikingly enough, the projects of integrated coastal development (recently also supported by the EU in a European demonstration programme) have hardly been discussed as paradigmatic for rural development and as indicative of development problems of rural areas that come with multiple resource use (Bruckmeier, 1998). The presently enforced opening of rural areas to non-agricultural resource users and multiple uses of natural resources may support the willingness to learn from development processes in coastal areas. From sociological or cultural-anthropological points of view, the differences between agricultural and coastal areas have been formulated by Mark

Nuttall as follows: by tradition, coastal areas are the outward-looking ones, where agriculture and rural development has always followed more complex forms of multiple natural resource use and interaction between resource user groups; inland rural areas are the more inward-looking ones where the world is a local one.[9] In both cases, different types of socio-cultural norms are of significance for the further development of such areas.

*Technological modernization* becomes significant for rural development when it is linked with strategies for social change that are accepted by the social and resource user groups and the residents of rural areas. Technological modernization that influences the rural areas in Europe happens presently through the building of 'trans-European networks' (using the political jargon of the European Commission). These networks are based on the integration of a super-infrastructure of material transport networks (roads, railways, water and air traffic) and immaterial infrastructure (of networked information technologies). However, the idea is deficient in the decisive respect that the Commission does not even think about social networks and cultural traditions as a precondition of successfully building the trans-European networks. These networks are only thought of as networks to achieve competitive advantages of European capital (Altvater and Mahnkopf, 1996, p. 455).

*Negative effects of agri-environmental policy* may become visible in various regions. In the worst case, this policy may – unintentionally – support the emergence of an 'inner periphery' of rural areas within the countries in the EU which can be seen as a reproduction of the North-South cleavage at regional levels. In most member countries a classification of rural areas is in use which identifies certain of these as peripheral and the structural and regional policy programmes of the EU emerge from this diagnosis. Are the rural peripheries regions where environmentally sound agricultural production will be found, whereas in the main agricultural areas intensification goes on? If this happens, rural development would become an example of the negative variant described above, where social and economic marginalization can turn into 'ecological marginalization' (Kousis, 1998). Ecological marginalization is the transformation of the countryside and rural areas to resource stocks and waste depositories under the influence of powerful external actors such as state or economic corporations – it is the transformation of rural areas to 'ecological backyards' which may exist as areas with overused and polluted resources as well as peripheral areas.[10] This can also happen under the influence of new agricultural technologies such as genetic engineering and the increasing role of non-food production in agriculture. It is not that this cleavage will separate the northern and southern member countries only – it can also become a regional segregation in every country.

# Notes

1 Data and information form the following research reports are used: Billaud et al., 1996; Billaud et al., 1997; Bruckmeier, Patricio, La Calle, 1997; Bruckmeier, 1999.

2 Most of these systems may be seen as specific modes of production, with the exception of traditional forms of agriculture which cannot be seen as production forms or ecologically adapted technologies only; they are parts of larger systems which require specific economic and socio-cultural institutions to become viable.

3 The discourses on regional, integrated and sustainable development tend to neglect established power relations. However, sustainable and regionally adapted development paths must give a realistic account for power structures, centre-periphery dependencies and unequal appropriation of resources and how such structures can be changed.

4 For details see Billaud et al., 1996.

5 It is difficult to account adequately for 'ecological services' as a by-product of agricultural production within a system of agricultural policy and property rights which is not built for the internalization of negative external effects at the level of the single production unit but does the reverse by stimulating intensification.

6 The Expert Council of Environment established by the German Federal Government has stated in 1994 that the participation of farmers in the agri-environmental measures is especially high in marginal land areas where extensive land use is already prevailing, thus also stating free rider effects of the regional programmes.

7 'The recent evolution of the CAP is ... encouraging the granting of subsidies which essentially represent support of owners of the larger holdings which, nowadays, as in former times, are the most powerful and protected social group in agriculture and rural society (Graca...) and whose position has been strengthened by the current crisis' (Baptista, 1995, p. 319).

8 Accounting for the negative external impacts of different types of agriculture seems to reveal a 'northern bias' of the agri-environmental discussion since it identifies as main environmental damages such resulting from modern intensive agriculture.

9 Both these trends towards integrated rural development and integrated coastal management are based on concepts which have been designed and applied before in development co-operation in developing countries. The policy instruments, the co-operation patterns and the networks that such integrated approaches require are more and more of the types of instruments that come into practice with the unfolding of the agri-environmental policy through the reform of the CAP: negotiative, persuasion oriented, participatory and consensus-based approaches.

10 The discussion of 'NIMBY regimes' (NIMBY: 'not in my backyard') in environmental policy illustrates the cases of shifting environmental pollution, overuse of natural resources or deposits of toxical waste etc. into areas (and countries) where the inhabitants, resource owners or political actors have not the influence and power to prevent the such inconsequent pollution management and shifting of the burden.

# References

Almas, R. (1998), 'Rural Development in the Norwegian Context', *Environnement & Société*, no. 20, pp. 79-85.

Alphandery, P. and Pinton, F. (1998), 'Le "Pays", Territoire de L'environnement?', *Environnement & Société*, no. 20, pp. 121-131.

Altvater, E. and Mahnkopf, B. (1998), *Grenzen der Globalisierung – Ökonomie, Ökologie und Politik in der Weltgesellschaft*, Westfälisches Dampfboot, Münster.

Baldock, D. and Lowe, P. (1996), 'The Development of European Agri-environmental Policy', in M. Whitby (ed.), *The European Environment and CAP Reform*, CAB International, Wallingford, pp. 8-25.

Baptista, F.O. (1995), 'Agriculture, Rural Society and the Land Question in Portugal', *Sociologia Ruralis*, Vol. XXXX(3), pp. 309-321.

Bazin, G. and Roux, B. (1995), 'Resistance to Marginalization in Mediterranean Rural Regions', *Sociologia Ruralis*, Vol. XXXV(3), pp. 335-347.

Bignal, E. and McCracken, D. (1996), 'The Ecological Resources of European Farmland', in M. Whitby (ed.), *The European Environment and CAP-Reform*, CAB International, Wallingford, pp. 26-42.

Billaud, J.-P. et al. (1997), 'Social Construction of the Rural Environment – Europe and Discourses in France, Germany and Portugal', in H. de Haan, B. Kasimis, M. Redclift (eds), *Sustainable Rural Development*, Ashgate, Aldershot.

Billaud, J.-P. et al. (1996), *Sociological Enquiry into the Conditions Required for the Success of the Supporting Environmental Measures Within the Reform of the Common Agricultural Policy*, Cologne and Brussels (Final Report EU Research Project No. EV5V-CT94-0372).

Bruckmeier, K. (1998), 'Integrated Coastal Zone Development in Human Ecological Perspective', University of Göteborg, *HERS-SUCOZOMA Report* no. 2.

Bruckmeier, K. (1999), 'Policy Influences on Agricultural and Livestock Systems in Different Regions of the EU: The Example of the Common Agricultural Policy (CAP) Reform's Agri-environmental Measures', in S. M. Williams and I. A. Wright (eds), ELPEN – *European Livestock Policy Evaluation Network, Proceedings of Two International Workshops*, Macaulay Land Use Research Institute, Aberdeen, pp. 93-106.

Bruckmeier, K. et al. (1997), 'A New North South Development Axis with the CAP Reform?', Paper presented at the Congress of the European Society for Rural Sociology, Greece.

Buckwell, A. (1997), *Towards a Common Agricultural and Rural Policy for Europe*, Directorate-General for Economic and Financial Affairs, Brussels.

Buller, H. et al. (eds) (2000), *Agri-environmental Policy in the European Union*. Avebury, Aldershot.

Cernea, M. (1991), *Putting People First. Sociological Variables in Rural Development*, 2nd ed., Oxford University Press & World Bank, New York.

Dent, J.B. and McGregor, M.J. (eds), *Rural and Farming Systems Analysis: European Perspectives*, CAB International, Wallingford.

Europäische Kommission (1985), 'Perspektiven für die Gemeinsame Agrarpolitik. Mitteilung der Kommission an den Rat und an das Parlament', Brüssel.

European Commission (1998), 'Agenda 2000 – the Legislative Proposals', Brussels.

European Commission, DG XII (1996), Global Environmental Change and Sustainable Development in Europe, Luxembourg.

European Commission, Directorate General for Agriculture (1999), Evaluation of Rural Development Programmes 2000-2006 Supported from the European Agricultural Guidance and Guarantee Fund – Guidelines. Brussels (VI/8865/99-Rev.).

Folke, C. et al. (1997), 'The Problem of Fit between Ecosystems and Institutions', Beijer Institute of Ecological Economics, *Beijer Discussion Paper Series* no. 108, Stockholm.

Garrido, F. and Moyano, E. (1996), 'The Response of the Member States: Spain', in M. Whitby (ed.), *The European Environment and CAP Reform*, CAB International, Wallingford, pp. 86-104.

Gaudemar, J.-P. de (1995), 'Macro-scénarios pour le Développement de L'arc Latin', in H. Karl and W. Henrichsmeyer (eds), *Regionalentwicklung im Prozess der Europäischen Integration*, Bonner Schriften zur Integration Europas, Bonn, pp. 131-164.

Hoggart, K. (1997), 'Rural Migration and Counterurbanization in the European Periphery: the Case of Andalucia', *Sociologia Ruralis*, Vol. 37(1), pp. 134-153.

Izcara Palacios, S.P. (1998), 'Farmers and the Implementation of the EU Nitrates Directive in Spain', *Sociologia Ruralis*, Vol. 38(2), pp. 146-162.

Jollivet, M. (ed.) (1997), *Vers un Rural Postindustriel. Rural et Environnement dans Huit Pays Européens*, L'Harmattan, (collection 'Environnement'), Paris.

Jollivet, M. and Eizner, N. (eds) (1996), *L'Europe et ses Campagnes*, Presses de Sciences PO, Paris.

Kousis, M. (1998), 'Ecological Marginalization in Rural Areas: Actors, Impacts, Responses', *Sociologia Ruralis*, Vol. 38(1), pp. 86-108.

La Calle Dominguez, J.J. and Velasco Arranz, A. (1997), 'Espagne – La Ruralité: un Concept Mort-né?' in M. Jollivet (ed.), *Vers un Rural Postindustriel*, L'Harmattan, (collection 'Environnement'), Paris, pp. 45-75.

Limousin, P. (1996), *Les Agricultures de L'Union Européenne*, Armand Colin, Paris.

Lourenco, N. et al. (1992), 'Imagens da Integracao: Representacoes Sociais sobre a Integracao da Agricultura Portuguesa na Comunidade Europeia', *Analise Social*, XXVII, nos. 4/5, pp. 955-971.

Moffatt, I. (1996), *Sustainable Development. Principles, Analysis and Policies*, The Parthenon Publishing Group, New York and London.

Moyano, E. (1995), 'Farmers' Unions and the Restructuring of European Agriculture', *Sociologia Ruralis*, Vol. XXXV(3), pp. 348-365.

O'Riordan, T. and Voisey, H. (1998), *The Transition to Sustainability: The Politics of Agenda 21 in Europe*, Earthscan Publications, London.

Pérez Yruela, M. (1995), 'Spanish Rural Society in Transition', *Sociologia Ruralis*, Vol. XXXV(3), pp. 276-296.

Ray, C. (1997), 'Toward a Theory of the Dialectic of Local Rural Development within the European Union', *Sociologia Ruralis*, Vol. 37(3), pp. 344-362.

Redclift, M. (1996), *Wasted – Counting the Costs of Global Consumption*, Earthscan Publications, London.

Rodrigo, I. (1998), 'Social Identities and Family Farming in Portugal: "The Old" and "The New" Countryside' in L. Granberg and I. Kovách (eds), *Actors on the Changing European Countryside*, Institute for Political Science of the Hungarian Academy of Science, Budapest, pp. 215-224.

Scheele, M. (1996), 'The Agri-environmental Measures in the Context of the CAP Reform', in M. Whitby (ed.), *The European Environment and CAP Reform*, CAB International, Wallingford, pp. 3-7.

Shucksmith, M. and Chapman, P. (1998), 'Rural Development and Social Exclusion', *Sociologia Ruralis*, Vol. 38(2), pp. 225-242.

SRU (1996), *Konzepte einer dauerhaft-umweltgerechten Nutzung ländlicher Räume. Sondergutachten*, Stuttgart.

Tronquart, J.-F. (ed.) (2001), 'Rural Development in the Context of the Agenda 2000, Mid-Term Review in the European Union and in the Applicant Countries' (*European Parliament, Directorate-General for Research, Working Paper AGRI 137 EN, Provisional Edition*, Luxembourg.

Volker, K. (1997), 'Local Commitment for Sustainable Rural Landscape Development', in *Agriculture, Ecosystems & Environment*, 63, pp. 107-120.

Waniek, R.W. (1995), 'Sektoraler und Raumstruktureller Wandel in Europa', in H. Karl and

W. Henrichsmeyer (eds), *Regionalentwicklung im Prozess der Europäischen Integration*, Bonner Schriften zur Integration Europas, Bonn, pp. 5-51.

Ward, N. (1993), 'The Agricultural Treadmill and The Rural Environment in the Post-Productivist Era', *Sociologia Ruralis*, Vol. XXXIII(3), pp. 348-364.

Whitby, M. (ed.) (1996), *The European Environment and CAP Reform*, CAB International, Wallingford.

WWF (ed.) (1994), *The Nature of Farming. Low Intensity Farming Systems in Nine European Countries*, Brussels.

WWF (ed.) (2001), *The Nature of Rural Development: National Reports and Synthesis Report.* http://www.panda.org/resources/programmes/epo/initiatiuves/agridev.cfm.

Chapter 5

# Changes in the CAP and the Future of Europe

Kostas Vergopoulos

## Introduction

The Treaty of Rome (1957), the European Union's (EU) constitutional chart, makes the economic and political unification of the old continent its target. The deep structural disparities or divergences that determine the European region, however, need to be rectified if the European economy is to function as a whole. Otherwise, the European project itself will remain under permanent threat.

Regional inequalities are already noticeable, especially in the living standards of Northern and Southern Europe. Similarly, disparities in the level and quality of social security services, employment and unemployment rates and comparative economic growth rates continue to exist. Consequently, in this context of multi-divergences, respective price systems among countries also continue to diverge: price levels are higher in the northern parts of the continent in comparison to price levels of the southern parts and in contrast to the levels of real incomes. The unemployment rate is only two per cent in the district of Rhineland, but it exceeds 33 per cent in Andalucia. Perhaps each zone in Europe is situated at a different historical momentum. If this is true, the problem of European unification proves to be much more complicated than once thought.

## Theories of Regional Integration

According to the general theory of economic integration, at both the regional and the international level, stability of the whole presupposes the adjustment of its components. But the real questions are what are the risks threatening the old continent's integration; and what mechanisms and implementation instruments are at the European economies' disposal for achieving adjustment and integration? Discrepancies in income, demand, levels of employment and unemployment have led to diametrically opposed imbalances between the North and the South of Europe – the risk of overheating and inflation in the European North and the risk of inefficiency and recession in the southern part of the continent. The organic weaknesses of the South can undermine the North's stability and conversely, overheating in the northern part can destabilize the southern region. Should the North endeavour to control the economy's overheating and inflation by implementing restrictive decelerating policies, it will not be securing against additional recessionary pressures coming from the South. In both cases, the degree of recession of the European system as a whole will be increased.

### The Concept of Economic Cohesion

Cohesion of the EU, in a context of stability, presupposes the existence of mechanisms that absorb the regional disparities or inequalities in the short term and simultaneously promote the structural adjustment in the long term. The European aggregate does not have as its sole aim international price competitiveness; it must also fend off the diametrically opposite risk – that of recession and price deflation, which is inevitably transmitted as a result of regional inequalities and unequal pressures on prices that have been witnessed repeatedly in Europe. European competitiveness is threatened by increased labour costs in the North. In turn European markets' stability is undermined by the exceptionally low labour costs in the South and the consequential demand and price inefficiencies from that region.

Traditionally, flexible economic adjustment is the result of unrestricted exchange rate floating and, therefore, one of prices expressed in terms of the floating currency. Given that priority is given to hard currency policies, however, countries inevitably have to resort to other adjustment strategies, such as compensatory mobility of the factors of production or institutional counterbalancing shifts of resources. As long as Andalucia does not send its unemployed to Rhineland or Castille; and German capital does not move to the South (or areas with medium growth rates) so as to take advantage of low labour costs; regional disparities will not be absorbed through spontaneous market functions. Inevitably, for Europe to achieve its targets, institutional compensatory resource transfers will have to be activated from surplus zones to deficit regions. Resource transfers will have to be even larger, given that they should counterbalance stiff exchange rates and inadequate production factor mobility within Europe. It is already well known that just as Europe's internal labour mobility remains at low levels, capital mobility does also.

*Cohesion amongst European Agrarian Sectors*

Since the 1960s, Europe has recognized the issues of stability, harmonization and integration in European markets. The elaboration of a Common Agricultural Policy (CAP), whose goal was the agricultural market integration of the member states to the European whole, was already evidence of the acknowledgement of this matter. In the end, the CAP has turned out to be nothing more than a system of stabilizing price intervention that aims to ensure the cohesion of the large internal European market and the structural reform necessary for long-term cohesion. European or national authority intervention concerning agricultural issues is considered inconceivable without the regular stabilizing shifts of resources that ensure the recycling of European surpluses to deficit regions. These mechanisms are enforced not only to ensure the harmonious functioning of Europe as its own entity, but also to secure the necessary structural reforms. In turn, the mechanism that absorbs agricultural inequalities presupposes political and financial solidarity among European state-members. Such a fact makes the necessity of political unification extremely urgent and, in all cases, impossible to be by-passed.

## Agricultural Market Stability and the CAP

For whatever reason, the CAP has remained the only existing common European policy. It has definitely succeeded in one achievement: the counterbalancing of political and fiscal disunity by channelling and redistributing direct funds, according to the need of agricultural areas (with priority given to stability) to achieve cohesion between all parts of Europe while ignoring the national sovereignty of each member state within the union. In the CAP model it is the agricultural regions, not the member states that deal with the European budget.

The stability of European agricultural markets is achieved not only when agricultural issues concerning member countries are taken into account but also when non-agricultural issues are taken into consideration. Even though the CAP exhibits many weaknesses in areas that should be reformed, experience has shown that economic stability and cohesion of the European whole cannot be attained without regular fund transfer between European regions. Fund transfers should be larger when production factor mobility is weak.

### New Course in Europe

In recent years, however, progressive, stable and radical change in European policy has come about resulting in the gradual abandonment of the goals of stability and cohesion in the large internal market. Foreshadowing this change were Britain's Former Prime Minister Margaret Thatcher's extreme proposals in 1984-8. In contradiction to every notion of stabilizing and cohesive European policy, the 'iron lady' enforced the renowned British order on its European partners. But the European partners legitimated Britain's arbitrary claims for the association of all European incoming and outgoing funds with national units. This changed the core of the policy determining European cohesion. Since then, national policy has come to predominate over common market policy, and national goals have come to be considered superior to the goal of European integration.

Toward the end of the 1980s, starting in Britain, liberal deregulatory policy catalysed European incorporation into the new liberal order of international trade through the gradual abolishment of any price regulatory systems. The CAP liberal reforms of 1992 introduced the principle of direct agricultural income subsidies to counterbalance the decrease in prices resulting from the disorganization of European agricultural markets. The final account, even though it was not as destructive as expected, given that incomes were kept at similar levels through subsidies, was unsettling because the process of consolidating European cohesion and integration was halted.

*From Reforming the CAP to the 'Subsidiarity Principle'*

Continuing appeals for further deregulation have now made the cost of the CAP a tremendous burden that increases the competitiveness of the European economy, rather than making it a necessary expense intended to promote stability within the European markets. Aiming to present the CAP as an excessively expensive policy, interested parties appeal to consumers' legitimate demands for cheap food products even if they come from distant markets that cannot be controlled. Generally speaking, pressures to abolish the CAP principles during the 1990s were based on claims of intensity of world market competitiveness, avoiding matters of stability and cohesion for the large internal market that had dominated common market discussions in the 1970s and 1980s.

With the Maastricht Treaty (1991), Europe established the 'subsidiarity principle'. In the name of this principle, in case of doubt, national jurisdictions take precedence over European jurisdictions, and Europe takes upon it responsibilities only when they have been specifically assigned. Since 1992, the viability of the CAP, and any concept of common policy have been small. Europe's support of agricultural prices has gradually been transferred to agricultural incomes. However, managing agriculture through the manipulation of agricultural incomes is not a measure equivalent to management via price manipulation. European agricultural prices are indicators of the unified European market; they therefore provide information on how European integration is progressing while the introduction of agricultural income criteria exacerbates the fragmentation of European market stability and cohesion. Income policy goals are not equivalent to those deriving from price policies: the stability of agricultural markets reveals the degree of European integration, while the maintenance of agricultural incomes reveals the gradual disorganization of European markets. From this point on, with the fixation on liberal economic policies, the idea of a united Europe is fading. Progressively it is being replaced with differing national situations based, in the case of agriculture, on the weight each specific country carries, and in the case of economic and political issues, on the circumstances in each country. The CAP' s liberal reform in 1992, which took place under the pressure of the United States and the World Trade Organization (WTO), enhances and strengthens national criteria and jurisdictions, working against the formation of an integrated and unified European agriculture.

**The EU Budget**

Pressures to dissolve the CAP are increasing, but new counterbalancing policies have been introduced over the past decade such as the structural European funds (regional, social, and development funds). Structural Funds are expanding, however, leaving less room for market stabilization funds. Structural Funds, to a degree, are meant to counterbalance the speedier shrinking of funds that were applied to market and price policies. Funds from both categories together do not outstrip the sums allocated to the very first CAP. So the most important outcome is that the expansion of Structural Funds in practice formalized the abandonment of the most important EU objectives: those of the stabilization and cohesion of European markets. The European budget derives funds equivalent to 1.14 per cent of European Gross National Product (GNP), a level that is still very low (theoretically it reaches 1.28 per cent of GNP). While in the past, the CAP funds reached 80 per cent of the Union budget, today total expenses for the countryside reach 80 per cent only if Structural Funds are taken into account.

The proposals contained in Agenda 2000 abolishing the CAP, have been renamed and termed the 'co-financing' and 'renationalization' of a larger proportion of European agricultural expenditure. Further price decreases have been introduced (with reference to international price levels) while the number of farmers being compensated on losses has also been decreased. Twenty-five per cent of direct income subsidies are to be nationalized. We should remember that 40 to 60 per cent of total effective support to agriculture is already based on national expenses, not European. With the proposed additional nationalization of direct subsidies, European agriculture is straying away from European integration and is being reinstated to its pre-integration national level. Today, in the context of globalization and deregulation, European agriculture is being 'de-europeanized' and reincorporated into a national framework.

This development is especially detrimental to the countries of Southern Europe. Given that national agricultural expenditure in these countries has been neglected, European expenditures represent ¾ of total support for Southern Europe's farmers. The consequences of the withdrawal of European agricultural funds are, therefore, much stronger in the South than in the North.

*'Net Contributor' Countries*

The CAP problem is not only working against stability and cohesion policies but is further validated by Germany and other wealthy European countries (Austria, Holland, Sweden) who claim that they desire to stop clear-cut contributions to the European budget in the existing framework. In order to stop this 'abnormality,' Germany and Austria proposed the nominal freezing of overall European expenditure (not only for agriculture). This will eventually result in the gradual decrease of contributions in real terms. The proposal to freeze general expenditure has already decreased the EU budget to 1.14 per cent of GNP and will shrink it to 0.9 per cent by the year 2006.

The current discussion concerning the CAP's future indirectly brings up two other very important issues: a) the requested balance of European transactions for each member country on a national basis and b) the freezing and shrinking of the community budget as a whole. It is obvious that all three issues lead to the same direction: fewer solutions on a European basis, more management on a national basis. If Chancellor Gerhard Schroeder does not want to 'squander German money' in Brussels, he could invent European principles that might prove favourable for Germany. Germany's net transaction balance with the EU could improve without dissolving common European policy and abolishing the objective of stability and cohesion of the great internal market. To date, national expenditure control policies were justified under the pretext of the historical decadence of member states. If in turn European expenditure is constricted, why should we visualize a different future for Europe than for each of its constituent member countries?

## Monetary Union and the National Equilibration Principle

Despite all of the above, Europe has succeeded in what was once unimaginable; it is moving towards monetary integration without prior political integration. (At the moment, political integration is not foreseeable in the near future). Chapters concerning common foreign and defense policies could represent a positive step towards political integration; however, these issues are at an embryonic stage. To date, what has been achieved is that major decisions on economic and monetary integration are taken by a group of sovereign countries. Decisions concerning the European Monetary Union (EMU) and the Euro continue to be taken not within the supranational context of economic and political integration but within a framework of rules that simply determine the co-operation of the fifteen sovereign member states. The prospect of a supranational context for the present is not faced, it is not even mentioned.

For a long time, economic and monetary unity was presented as the linchpin that would force Europe to follow the path of political integration. Today we have come to realize that the reverse is true. Monetary unity is promoted on the one hand; European desire and political goodwill is absent, undermining the prospect of monetary union. It is possible that during a third phase we will be in a position to come to conclusions that are more balanced. On the ground of regional integration among most of the sovereign member states and their peoples there can be no secret linchpins. The process of integration ought to be based on outspoken political goodwill that is often renewed in a democratic manner.

## The Concept of Convergence

Suspicious attitudes towards the prospect of political integration make monetary integration uncertain and problematic. Today Europe is not pursuing cohesion and convergence of its economic space; instead, it is pursuing the simple convergence of monetary indicators that determine national macroeconomic entities. The German convergence model of today differs from that of a cohesive European structure as it was discussed in previous decades. The concept of cohesion and integration requires the mobility of resources, periodic interregional transfers of funds and the predominance of European institutions and statutory rules as determined by the CAP model. Today's convergence model has put the burdens and costs of necessary adaptation on the shoulders of each member state without

European resources and therefore without regional transfers or the predominance of European institutions and management. In the past, the aim was for the cohesion of Europe as a whole; whereas today the notion of convergence does not encompass Europe as such but the fifteen national subsets.

The prospect of European expansion towards the East should be regarded as an opportunity that will strengthen the continent's cohesion. However, today, given that Germany is giving priority to national policies, expansion towards the East is viewed as an opportunity to shrink European expenditure further. When Europe freezes its expenses, it really freezes the implementation instruments and mechanisms necessary for its own deepening and maturity. Obviously, nothing can be considered definite forever. The proposed reforms shift the weight of adaptation onto the weaker partners, who in this case remain helpless and will be exposed to penalties should they violate the restrictive rules. This fact makes the adaptation of the whole an irresolvable problem. Today Europe is speedily drifting away from the principle of financial solidarity between its members – a principle determined in the Treaty of Rome. In practice, Europe is shifting away from its own principle of stability and cohesion.

## From National Co-financing to Restoration of National Policies

The contours outlining the Union's prospects have already been announced by Germany's Minister of Finance: no taxation alignment and no tax integration in a European context; instead, simple co-ordination of member states. Likewise, Germany's former Minister Oscar Lafonten declared that unemployment could only be challenged on a pan-European scale based on European programmes and means. Despite these declarations, Gerhard Schroeder's Government has already accepted the opposing principle; the negotiation of social contracts between German employers and trade unions solely on a German scale. If German trade unions have been assured that transfers of resources to the South do away with jobs in the North, Spanish unions have been assured that the opening of internal economic borders has been one of the most important causes of unemployment in the South. Even though both sides agree that only Europe can give solutions where member states cannot, concerns still exist about the point at which member states 'nationalize' their solutions, leaving the rest for Europe to solve. In such a case, in the near future there will be neither Europe nor 'national' solutions. Is the gap between Northern and Southern Europe still so deep and insurmountable that Europe must be sacrificed?

The French Agricultural Minister, in opposing the turnaround the Germans have made, condemned Germany not because it arbitrarily and one-sidedly transformed the very nature of European integration but because Germany 'intends to solve its problems inflicting the weight of the problem on France.' If the principles of financial solidarity among member states and European cohesion are abandoned, however, the implications are serious for European agricultural integration; the nature of future relations between France and Germany; the relationships between North and South; and most of all the character of the European structure.

On the other hand, opposing German policies of 'renationalization' and 'co-financing', the French position can be summarized as follows: the 'declining and gradual' shrinking of agricultural subsidies, having as its sole criterion the height of family incomes. Finally, the French proposal does not really turn against the German but reflects the same spirit of 'European defeatism' and abandonment characteristic of the German position. The Berlin Summit meeting of March 1999 did not reinstate the abandoned issues of European stability and cohesion, nor those concerning the organization and management of internal markets; instead it distributed the costs that derive from the gradual extinguishing of the CAP to farm households. The concept of declining and gradual shrinking refers to the distribution of cost (a result of predetermined shrinking) to groups of agricultural income in time. At the same time the need to decrease the CAP in general becomes formal policy. If the German position, through proposals of 'renationalization,' directly abolishes European policy and management, French proposals come to the same conclusion in time; no European policy will be possible in the future given that the shrinking of European expenditure has already been decided.

## Convergence Against Cohesion

Neither Lionel Jospen nor Gerhard Schroeder misses an opportunity to remind us that Europe's main market is its internal market, given that the EU exports only 8 per cent of its GNP. If such verification leads to the conclusion that the European framework should be strengthened and its regulatory role be enhanced within the internal market, Europe could profit from moves towards strength, cohesion, character and depth. If, in opposition, such verification leads Europe to the conclusion that more internationalization and deregulation is necessary, then the European structure will weaken as a result of the permanent lack of resources, institutions and policies.

As long as Europe does not recognize its responsibility to maintain stability and cohesion of internal markets and challenge unemployment, it sells out two decisive issues that legitimize Europe in public opinion. Convergence models, in current conditions, formed to suit arguments in support of international market integration, in substance destroy the prospect of European integration. The immediate consequence is that in the resulting gap formed, the old national complementarities, (already considered outmoded) as well as previous neighbouring relationships, divergences, inequalities and antagonisms among states not only on economic and political issues but social and cultural issues will reposition themselves. Given this prospect, true dangers lie ahead not because of historical innovation and the uncertain course followed thereafter but due to exactly the opposite reason: the lack of any innovation and the repetition of realities and situations already known from the past.

# References

Blogowski, A. (1996), 'Evolution du Financement Communautaire des Marchés', *Notes et Etudes Economiques*, no. 1.

Blogowski, A. (1996), 'Les Dépenses Agricoles de L'Union Européenne', *Notes et Etudes Economiques*, no. 2.

Bonnet, A. et al. (1996), 'La PAC et les Transferts entre Agricultures de la CEE', *Notes et Etudes Economiques*, no. 2.

Brinbaum, D. (1995), 'La Réforme de la PAC', *Revue Paysans*.

Chambres D'Agriculture (1996), *Revue* 1994/4, nos. 9 and 12.

Hairy, D. et al. (1996), 'Evolution et Recomposition des Concours Publics à L'Agriculture Depuis la Réforme de la PAC', *Notes et Etudes Economiques*, no. 1.

INRA (1992), *Réforme de la PAC*, Paris.

Korakas, (1998), 'L'Avenir de L'Agriculture Européenne', *Economicos Tachydromos*, (in Greek).

Krugman, P. and Dehasa, G. (1992), *EMU and the Regions*, Group of Thirty, Washington.

Ministere de l'Agirculture (1996), *Les Concours Publics à L'Agriculture, 1991-1995*, Paris.

OECD/OCDE (1996), *Politiques,Marchés et Echanges Agricoles*, Paris.

OECD/OCDE (1996), *Structural Indicators*, Paris.

Pisani, E. and Gousios, D. (1997), 'L'Avenir de L'Agriculture Européenne', *Economicos Tachydromos*, (in Greek).

Union Europeenne (1995-96), *La Situation de L'Agriculture dans L'Union Européenne*, Bruxellles.

Union Europeenne (1995-96), *Rapport Général D'Activité*, Bruxelles.

# PART II

# THE PROSPECTS OF RURAL DEVELOPMENT IN SOUTHERN EUROPE

Chapter 6

# From Common Problems to a New Policy Architecture: a Portuguese Perspective on Mediterranean Rural Development

José Portela and Chris Gerry

## Introduction

The purpose of this chapter[1] is to foster discussion about rural development in the Mediterranean context[2] and, with this aim in mind, we take as our point of departure two fundamental assumptions concerning rural economy and society. The first is a matter of empirical fact: agriculture *still* constitutes the vital core of the rural economy, particularly in the least favoured areas of the Mediterranean countries of the European Union (EU), though undoubtedly the centre of gravity of the rural economy has shifted (Portela, 1994c; Portela, 1999b). Farming is nowadays increasingly undertaken in combination with other activities such as rural tourism, local handicraft production, hunting and environmentally-linked recreational activities.[3] Needless to say, farming and forestry play a key role, currently occupying approximately three-quarters of the combined area of the EU's fifteen member states (CCE, 1993). The majority of analysts and policy-makers now recognize – or at least their discourse suggests as much – that agriculture is much more than just a specific economic activity and that, to have a future, it has to be liberated from the narrow, abstract and technicist approaches of the past, or from what Massot (1998: II.8) calls:

> typically sectoral, archaically productivist policies that emphasize the quantity of output rather than rural people and the territory that sustains them; designed – as it was – with the six founding member states in mind, it is now, in the final analysis, internally unbalanced (our translation).

The second point is more a question of analytical options. We take it for granted that the current malaise of agriculture and the design of more appropriate policies for its future development can only be fruitfully addressed if problematized from a combined socio-cultural, economic and environmental perspective (Nijkamp, 1990). Thus, 'agriculture is primarily a societal issue' (Portela, 1994a, p. 48) because, in many respects, rural life is the product of three fundamental interacting forces: the natural environment, economic endeavour and what we might term socio-cultural imperatives. Culture, argues van der Ploeg (1992, pp. 35-36), is located at the interstices and in the interaction,

> between internal and external relations, between experience and perspective, between past, present and future. Culture is not a phenomenon "outside" the so-called 'hard realities of market and technology' [...] [it] is not to be eliminated from the analysis, nor from the (theoretical) representation of agriculture. Culture is at the heart of it.

We set the stage for the discussion of Mediterranean rural development by first stressing that not only does Northern European farming differ considerably from that of the Mediterranean, but also the contexts in which they are embedded are quite dissimilar. We then advance the idea that, within Mediterranean agriculture as a whole, there may be more problems in common than is normally admitted; this implies that there may also exist greater room for collective or integrated manoeuvre. These views also take into account both the common features and the specificities of the development path followed by the Central and Eastern European Countries (CEECs) that are knocking at the EU's door.[4] Assuming that it will be open, the following section tries to go beyond the over-cautious pragmatism of mainstream proposals for a reformed Common Agricultural Policy (CAP) by sketching out four principles for a more radical redesign, along with three options for a new policy architecture. Finally, we argue that in spite of the difficulties inherent in satisfactorily defining agriculture in a manner that transcends prevailing economism and productivism, agriculture nevertheless still has to be the starting point for achieving meaningful and sustainable rural development.

## Distinctions between Mediterranean and Northern Farming

Everyone accepts that the physical environment, i.e., relief and climate, varies greatly within Europe; indeed, this variation helps to distinguish Northern from Southern Europe, and Northern from Southern European farming. Clearly, once we focus down from this general and abstract dichotomy to the specificities of the national, regional and even local level, the picture is less dualistic, much more complex and differentiated (Portela, 1988). Clearly, in every southern rural reality there are northern enclaves and, even in the North, much that appears southern still subsists.

With reference to some of the key distinguishing features of the agrarian economy and society in Europe that we also use in the present chapter, Oliveira Baptista identifies a clear North-South divide:

> We can identify two clearly differentiated and predominant types of agricultures in the contemporary EU, [...] Geographically, one type is limited to the North of Europe

(Denmark, Sweden, Finland, Holland, Belgium, Luxembourg, The UK, Germany, Austria and France) where returns per unit of labour are high, and the few people who work – mainly full-time – in agriculture are highly professionalized, and the other corresponds to the South (Italy, Greece, Spain and Portugal), where the reverse is the case. The sole exception is Ireland which, despite it high levels of professionalization in farming, in all other respects reproduces the features of southern countries (our translation) (Baptista, 2001, p. 63).

At least as important as the physical features of Northern and Mediterranean Europe are the differences in their respective histories – both in general terms and, more specifically, the history of their agriculture – with particular regard to the relations between farming (on the one hand) and the State, political and agricultural producer organizations (on the other). It is worth remembering that most Mediterranean countries – namely, Greece, Italy, Spain and Portugal – have lived under dictatorships for a considerable part of their recent history, and these corporatist regimes imposed lasting perverse effects on the economy and society alike.

In Portugal, the Salazar dictatorship left deep marks on State structures, which still tend to be more rigid, static, centralized and addicted to top-down approaches than is the case in conventional state bureaucracies. More specifically, the inertia and lack of innovation which characterizes the parts of the civil service that have direct or indirect responsibility for agricultural issues, exacerbates the negative effects of already highly complex CAP bureaucratic procedures. As a rule, the State has a poor record of meeting its stated commitments to agriculture: as a result, farmers' attitudes towards officials are at best sceptical, at worst openly suspicious (Portela, 1988). Contacts between them are cautious and reserved, and farmers tend to 'dance according to the music': they welcome benefits, lobby for better output prices, reductions in taxes and key input prices, and try to avoid problems, as everyone does.

We should keep in mind that in the case of Portuguese agriculture, farmers' organizations were an integral part of the corporatist State, and were controlled through the local rural elite. Those organizations were deprived of autonomy, and farmers' participation was reduced to a compulsory formality. Despite major political and institutional change since the 1974 restoration of democracy in Portugal, farmers' organizations are still very weak, both financially and politically; the attitudes and practice of many current members still seem to reflect the old order and, as a result, their real participation in the organization's life and in local development initiatives is severely inhibited (Portela and Cristóvão, 1991). Inter-related with such behavioural and attitudinal factors, a number of severe structural and institutional constraints limit the extent to which farmers are able to influence decisions at *any* level, from their village or region to Lisbon, let alone Brussels. In brief, differences in social organization and leadership in farming and in rural society have set Northern and Southern European countries apart.

Broadly speaking, southern farmers may also differ from their northern counterparts in two further, interrelated ways: most of them have not proceeded very far down the path of orthodox agricultural modernization, that is, they have not experienced the decades of productivist agriculture[5] that have come to characterize the North, thus remaining unaware, to a certain degree, that there exists such a trap into which they may fall. This leads us to conclude that Mediterranean farmers may be inherently somewhat less conscious of the negative environmental impacts of productivist policies, for three main reasons:

1. As a rule, in the South, the main sources of pollution are related to heavily urbanized and industrialized areas, to tourism, and to intensive factory farming;
2. Farmers legitimately wish to improve their incomes and livelihood and are highly conditioned to see that 'modernization' of their productive structures and agricultural practices is the only way to achieve that goal; and
3. Farmers are not stimulated either by the CAP or by the market to keep to their 'traditional', low-input farming.

On the other hand, southern public opinion and consumers may be just as sensitive to the issues of the environment as their northern counterparts, particularly with regard to the health implications of consuming foodstuffs produced and processed under 'modern' conditions. However, the fact that they do not have comparable incomes means that their propensity to demand 'green' food may be lower. Paradoxically, Mediterranean farmers – albeit unwittingly – may be more 'post-modern' than their northern counterparts. Barely conscious of the extent to which their products are seen as a benchmark for improving the quality of *northern* agriculture, they have yet to reap the financial rewards of their still largely unpolluted resources (soil, water and air) and their traditionally more environmentally-friendly farming methods.

History also helps to explain the marked differences in the size distribution of holdings and ownership/tenurial patterns in the two areas.[6] Broadly speaking, in Northern Europe land holdings are relatively large, while in the South, small-holdings predominate. According to Eurostat, the average farm size in circa 1990 was approximately four hectares in Greece, five in Portugal, six in Italy and fourteen in Spain, while in Ireland, Denmark and the United Kingdom the corresponding figures were twenty-three, thirty-two, and sixty-four hectares respectively. Moreover, farm plots are fragmented and scattered in the South, more concentrated and consolidated in the North. Also, in the period 1965 to 1990, the northern pattern became increasingly pronounced, while in the South there were few fundamental changes, either in the structure of ownership or in the production system (CCE, 1991). In the same period, the average farm size doubled in Belgium and Luxembourg, with increases in Holland, Germany and Denmark of between 60 per cent and 70 per cent.

Although family labour continues to be used to a greater or lesser extent in agriculture throughout Europe, in the South, neither farm wage-labour nor agrarian capital play the prominent and direct role that they do in the North. It has been the effort of family members, rather than the use of a variety of specialized farm machinery, that over the years has left its mark on the agrarian system. In Northern Europe, many rural women work in towns, but in the Mediterranean area, rural women are still particularly active in farming. Moreover, both in Portugal and in

Italy, women's participation in farm work is on the increase (CE, 1994, p. 20). The agrarian labour force in the South differs greatly in other respects, too: a great number of male family members are pluriactive, resorting to unskilled, complementary low-wage work or self-employment in parallel with their farming activities; indeed, the combination of farming with off-farm work (both local and non-local) is a key strategy for economic survival (Baptista and Portela, 1995). Nevertheless, pluriactivity also constitutes a major source of farmers' vulnerability, since multiple jobholders are excluded, more often than not, from national and EU aid. Thus, it is no wonder that the CAP, a northern-centred policy, has been unable to prevent the massive out-migration that has taken place in the continent's designated 'less favoured' areas, which are relatively more heavily concentrated and numerous in the Mediterranean countries.[7]

Broadly speaking, northern agriculture has somewhat monolithic, homogeneous patterns of land-use and highly uniform farming practices, whereas, in the South, these are much more extensive and diverse. The predominance of flat or gently rolling arable land in the North, used for cereals, and large-scale intensive milk and meat production is in stark contrast to the terraces and/or steep slopes of the South, planted with vineyards, orchards and chestnut, olive and almond groves. Not only are the typical Mediterranean products different from those characterizing northern farming, but also the scale and intensity of production is distinct: low intensity agriculture predominates in the countries of Southern Europe, not only in terms of the wider range of such farming systems to be found, but also the proportion of total agricultural land they occupy. Indeed, as Pinheiro remarked:

> Portugal, Spain and Greece all have 60 per cent or more of their farm land under less intensive systems; the area of such farms in the Iberian Peninsula makes up approximately 50 per cent of the total in the nine countries studied (our translation) (Pinheiro, 1997, p. 196).[8]

When charged with blindness to the distinctions between Northern and Southern European farming systems, the technical-administrative elite of the European Commission tends to argue that no such dichotomy exists, since beef, milk, as well as all sorts of arable and fruit crops are produced across the length and breadth of the Union. As the statement stands, no observer of Portuguese, Spanish, Italian or Greek agriculture could deny it; however, as Massot (1998, II.8) points out 'it is equally indisputable that both the degree of specialization and level of productivity of each member state's agriculture are extremely different' (our translation).

In the South, the soil is often poor and of low productivity, subject either to excessive rainfall or to drought (or even both), and often susceptible to erosion. The still pervasive itinerant system by which small flocks of sheep and goats are raised under the shepherd's watchful eye symbolizes a key characteristic of southern farming which is so often neglected – namely the harsh living and working conditions, largely unmitigated by either appropriate technology, adequate transport or decent housing. However, in comparison with the North, the rural areas of these countries are also endowed with classic agrarian landscapes, rich bio-diversity and generally good air and water quality. However, being only minimally commoditized, the potential value of these public goods enters even less into the everyday economic calculus in the South than in the North.

Over the centuries, southern farming has proven itself capable of adapting to the obstacles thrown up by a milieu that is hostile in the physical, economic and socio-political senses (Marques and Portela, 1994). Of course, there are objective reasons for the declining interest in agriculture and the very low social prestige that southern farmers 'enjoy'. The low returns on agricultural activity constitute one such reason; indeed, low farm incomes reflect many of the physical difficulties and socio-political disparities facing Mediterranean agriculture. Farmers in well-endowed areas, such as the Northern European plains, may derive incomes from their activities several times those earned in the South. A recent study undertaken for the European Commission (DG XI), recognized that 'mountain agriculture income remains two to four times less than income in the lowland regions, a large part of which comes from the distribution of public aid' (Euromontana, 1997, p. 24). However, it should be noted that not only do the Mediterranean countries have lower farming incomes, but the income differences between their more-favoured and less favoured rural areas are negligible (CCE, 1993, p. 25).

The way in which subsidies are distributed is a key factor in the wide intra-EU disparities in farm incomes which, in turn, constitute a serious obstacle to social cohesion. In Portugal, Spain, Greece and Italy, the average amount of subsidies per farm, less favoured areas included, is markedly lower than in other member states, due to smaller farm size and the lower unit level of subsidy paid. Even in less favoured areas, higher levels of support are enjoyed by the northern countries. Put succinctly, subsidies provided in the South, where the incomes are lower, are often lower than those provided in the North, where farming incomes are much higher (CCE, 1993, p. 28).

The following comparison underlines the contradiction in the way in which the aim of social cohesion is supposedly being pursued via the CAP. At the beginning of the 1990s, the share of subsidy in farm income in less favoured regions of the United Kingdom, France, and Germany amounted to 88 per cent, 69 per cent and 56 per cent respectively, while the corresponding figures for Spain, Italy, Greece and Portugal were respectively 13 per cent, 15 per cent, 30 per cent, 44 per cent (CCE, 1993, p. 29). The average ECU subsidy per farm varies enormously from one less favoured area to another: it amounts to about 18.250 in the United Kingdom, 10.900 in Belgium, 8.700 in France, and 7.700 in Germany, compared to a paltry 850 in Spain, 1.360 in Italy, 1.520 in Portugal and 2.150 in Greece (CCE, 1993, p. 29). This eloquently demonstrates how discriminatory the common policy has been, and how little it has done to meet common interests, let alone promote social cohesion.

Most Mediterranean crops have been excluded from the CAP income aid and since this is calculated on the basis of 'historical productivity', those most in need of improving their performance get least assistance. Thus, it is not surprising that countries such as France, Germany and the UK are the main beneficiaries of the CAP. In fact, together they receive one-half of EU agricultural expenditure, while representing only one-fourth of the number of farms and under one-third of the labour employed. In this regard, the distinct and discriminatory way in which the most recent Agenda 2000 reform of the CAP treats Northern and Mediterranean farm products, provides little if any comfort to southern farmers. Unequal treatment carries through into the sphere of agricultural research, too: even the policy reform impact assessments recently commissioned by the EU have tended to focus on the predominantly northern cereals, dairy products and beef, rather than on other sectors in which the Mediterranean countries specialize.[9]

While, in general, the level of commitment of both government and public services, as well as the quality of agriculture-related information systems are very uneven from country to country, all tend to be lower in the South. Thus Mediterranean farmers suffer not only from the biases inherent in the overall policy design, but also from the perverse effects of policy implementation resulting from the deficiencies of the state apparatus in their own countries.

The national budget plays a key role with regard to the 'indirect' effect of compensatory allowances, since member states are only partly reimbursed (within the limits of the established co-funding rates), and this is done *a posteriori*. The practical application of the principal of co-funding, and the level at which co-funding rates are set, encourages southern governments to reduce the level of the premium per farm unit. In short, farmers do not get the maximum sum that would be theoretically possible on the basis of EU regulations (CCE, 1993, p. 38). These regulations assume that member governments provide the full 'top-up' i.e., transfer to farmers the maximum co-funding component of compensatory payments. This may *not* be the case, especially when government expenditure is under constraint, whether due to national or EU policy imperatives.

Furthermore, it is in the 'problem' rural areas of the Mediterranean countries, that the asymmetries outlined above are most pronounced. There, the proportion of farmers in those less favoured areas eligible for payments in compensation for natural handicaps is lower than in other countries of the EU. The figures show how an ostensibly 'common' political space, in reality, may be divided by quite distinct frontiers. At the beginning of the 1990s, in the less favoured zones of Ireland, Belgium, Germany and the United Kingdom, the number of farms benefiting from the compensatory payments scheme amounted respectively to 77 per cent, 66 per cent, 59 per cent and 59 per cent; the corresponding figures for Greece, Portugal, Spain and Italy were 36 per cent, 22 per cent, 17 per cent and 10 per cent (CCE, 1993, p. 35).

If we focus our attention more specifically on the recent evolution of the less favoured zones, the situation becomes even more perplexing. Indeed, less favoured areas constitute a surprisingly large proportion of the total area of some northern countries: in 1990, they made up 99 per cent of Luxembourg, 54 per cent of Germany, 53 per cent of the United Kingdom, and 45 per cent of France (CCE, 1993, p. 40). Even more astonishing is the fact that, despite almost two decades of policies and the undoubted improvements that have been made, this proportion has *increased* rather than diminished. This is particularly noticeable in Germany, where the proportion of total hectares designated as less favoured areas was a third in 1984, but had risen to 54 per cent by 1990; equally, in France, this figure rose from 37 per cent to 45 per cent in the same period (CCE, 1993, p. 40). Rather than representing a higher real incidence of characteristics associated with less favoured status,[10] what has happened is that, quite simply, areas that were considered 'normal' have been redefined as less favoured, in order to reap the respective multiple advantages, namely compensatory payments, higher rates of co-financing of investments, increased sheep and goat premia, etc. In contrast to this situation, less favoured areas have hardly expanded in Italy, Portugal, or Spain, though in Greece, the area covered raised from 68 per cent in 1984 to 78 per cent in 1990.

Thus, in many respects, the CAP has exacerbated the specific – or 'natural', some might claim – distinctions between northern and southern agriculture. Indeed, the disparities between northern and southern agricultural incomes have tended to widen, due to the distinctive geographies and histories of Mediterranean member states of the EU and, in particular, the specificities of their agriculture, and their more limited ability to influence policy design and implement the CAP provisions to their own advantage. Policy makers have been reluctant to admit the extent of the failure of past economic, structural and welfare policies to generate socially and environmentally sustainable material progress in the countryside. Furthermore, it is only really at the level of discourse that the plans for a revamped CAP have begun to explicitly take into account the fact that different actors in rural society, public institutions and the agro-industrial filière have quite varying abilities to adapt to the structural changes Europe has undergone over the last fifteen to twenty years. There clearly still exists an enormous gap to be bridged between the EU's discourse of economic convergence and social cohesion, and the lamentably inequitable situation described above.

### Key Non-Farming Differences between Northern and Southern Europe

Agriculture does not exist in isolation. The broader economic, political and social context strongly conditions its functioning and performance. If southern agriculture is itself quite distinct from that practised in the North, then these differences are all the more marked when we examine how farming articulates itself with its 'external environment'. In our view, the socio-economic viability of the rural areas crucially depends on a number of key interfaces established between farming and

1.  The non farming business environment, i.e., the upstream and downstream components of industrial and service value-chains (including those into which farmers may have diversified);
2.  The networks and facilities for accessing information and knowledge;
3.  The physical and telecommunications infrastructures that links it to the wider economy and society; and
4.  The welfare regime that complements agricultural incomes.

### *The Non-farming Business Environment*

The industrialization experience of Northern European economies has been quite distinct from that of Mediterranean ones, the latter sharing a set of essential features which, taken together, constitute what has been variously referred to as 'intermediate', 'late' or 'dependent' industrialization. Their specific development trajectory has caused the economies of the South to be attributed a quite specific and restricted role, both in the EU and international divisions of labour, from which it has proven rather difficult to extricate themselves. Even after their admission to the EU, the relations of structural interdependency of Portugal, Spain and Greece (on the one hand) with the economies of Northern Europe (on the other) has tended to remain broadly the same: the South buys from the North much of the capital goods and other sophisticated industrial products it needs, providing the North in return with agricultural goods, and the less elaborate (albeit varied) output of its manufacturing industry, including consumer durables. Despite the impressive expenditures made in recent years on basic research, Research And Development (R&D) and the expansion of higher education, Southern European countries still remain technologically dependent,[11] particularly though not exclusively on their Northern EU partners. The persistence – or even deepening – of this unequal relationship has made it particularly difficult for the Southern European economies to generate and/or retain value-added in the same way and at the same rates as those of the North. It has also made them particularly vulnerable to the effects of economic downturns in the core economies (e.g., the crises of 1979-83 and 1990-92), and to the demands of competitiveness, adjustment and economic restructuring, whether imposed by economic crises, globalization and/or by the demands of European convergence.

While nominal macro-economic convergence was achieved in the 1990s by Portugal and, with somewhat greater difficulty, by France, Spain and Italy, real convergence with both the industrial structure and patterns of consumption of the North still has a long way to go. Notwithstanding the ostensible advantages of the digital economy, the increasing concentration of market power resulting from the late 1990s wave of mergers and acquisitions in the banking and financial sectors, will tend to restrict rather than enhance the capacity of Southern European enterprises to access the financial resources necessary to converge more rapidly with conditions in the North. It appears that, with the exception of the few that will succeed in competing with northern firms, and those that disappear altogether, southern non-agricultural enterprises will have to accept the inevitable: direct or indirect 'absorption' into filières dominated by firms based in Northern Europe or beyond.

Perhaps of most relevance to the rural areas is the probability that such firms will experience a progressive loss of entrepreneurial autonomy – for example over the characteristics of their product range and the technology used – resulting in their reduced 'territorial embeddedness', as well as limiting their capacity to transform the value-added they are able to retain into positive local development outcomes.

*Networks and Facilities for Accessing Information and Knowledge*

Over the years, northern countries have constructed a relatively effective knowledge chain, linking R&D, education, training and application, the effects of which are felt across all sectors. However, in the Mediterranean countries, a combination of indifference to science in general and insufficient investment in scientific education in particular, have led to a reliance on imported productivist solutions without sufficient attention being paid to their applicability to local conditions. Many of the problems of Mediterranean farming can be explained by gaps in this knowledge chain, resulting primarily from attitudinal and institutional constraints on both research and extension (Portela and Cristóvão, 1991). Researchers are often subjected to conceptual and methodological reductionism as well as alienated from farmers' real problems (Portela, 1994c; Portela, 1994d), a clear illustration of which is the persistent undervaluation of *traditional knowledge*, despite its importance in promoting sustainable agriculture (Portela, 1994b; Dries and Portela, 1994). Moreover, when new and relevant ideas and results emerge 'from above', their diffusion to farmers 'below' often fails to take place. As Cristovão et al. concluded:

> a major challenge lies ahead: recognizing that extension work requires a definite move from 'planning for' to 'planning and creating with' and finding the ways that may help reach this change in each context (Cristovão et al., 1997, p. 64).

The process of agricultural modernization tends to be driven more by supply conditions than the needs of local farmers. The priorities of the external productive, commercial, financial and policy environment, mediated through local public and private knowledge intermediaries, exert a decisive influence on farmers' investment decisions, promoting greater structural dependence and homogenization of Mediterranean agriculture, and may even threaten the viability and sustainability of many of the investments made by southern farmers (Portela 1994b, p. 277). Even where farming activities are relatively profitable, upstream input suppliers and downstream wholesale and retail firms tend to control key specialist knowledge, thereby limiting not only the income and investment potential of the more successful farmers, but also their capacity to internalize the technical and marketing knowledge that might boost both their own and their locality's development.[12]

In Northern Europe, the importance of information and knowledge to the success of farmers' initiatives is widely recognized. Business performance, employment creation and value-added retention can all benefit if the knowledge chain is developed in line with local needs and priorities: farmers' purchasing, production and investment decision-making can be improved if information regarding weather and markets is more accurate and more quickly disseminated, if appropriate specialist skills are channelled into advertising, marketing and distribution initiatives, and if competent and confident local institutions, capable of pooling knowledge, are encouraged.[13]

This is likely to be a slow process in many Mediterranean countries, where the vagaries of climate, policy[14] and/or entrenched rural clientelism can so easily undermine success. Furthermore, while the state channels for disseminating new practices have shrunk considerably due to the bureaucratic overload imposed by the CAP and its various revisions and reforms, the local associations that were intended to provide services in substitution for the state often do not have the requisite resources or skills.

Thus, key links in the knowledge chain are either absent, weak or incipient, and the impact of this structural context cannot be ignored. It is therefore hardly surprising that rural youth leave the rural areas, since they are confronted by agriculture that science has failed to understand and improve, and an educational system whose values and management are biased towards the city and the wealthy. The most able youngsters see secondary and higher education as the principle means of escape; for the rest, the attempt to complete compulsory education is simply drudgery, secondary drop-out rates are high, and local employment opportunities – where they exist – far from alluring (Portela et al., 2000).

*Physical and Telecommunications Infrastructures*

The rural areas of Southern Europe are lagging seriously behind their counterparts in the North, both in terms of the 'hard' and 'soft' aspects of development – respectively, transport and related infrastructures and access to the new telecommunications and information technology. In this regard, national governments cannot afford to rely exclusively on initiatives and funding from Brussels, but must intervene actively and forcefully to provide not merely the 'enabling environment', but also the concrete means for the rural economy to catch up and compete more successfully in Europe. In this regard, it is essential to find effective means to improve farming's capacity to adapt to and share in the benefits of the new transport and telecommunications infrastructures on which its future partly depends.

It is almost a cliché to argue that modern transport infrastructures, as well as improving both physical and market access of local residents, may have the perverse effect of *draining* rural areas both of the lion's share of value-added created there, as well as key segments of the economically active population. In part, this danger may be reduced if local decision-makers are able and willing to influence the priorities set for their regions. For example, since its accession, a large part of Portugal's so-called 'regional' development funding has been spent on 'national' and 'urban' development: infrastructure projects, particularly improvements to the motorway and arterial road network and urban beautification[15] (Portela, 2001). More has to be done to improve minor roads so as to facilitate the working of dispersed farm plots, to promote specific land consolidation projects,[16] or to open areas of outstanding beauty to visitors in a way that encourages local employment, endogenous products and environmental conservation.

Local strategies must be developed that can at least mitigate the potential damage caused by the 'opening up' of the rural areas. Ultimately their success will depend on liberating the potential of local human resources, creating incentives both to fix the local population that may be otherwise induced to emigrate, as well as attracting 'new rurals' if possible, along with determined support for initiatives that may combat the outflow of value added. If central and local government, as well as producers and their associations, fail to appreciate that modern infrastructure and new technology does not automatically create jobs *everywhere*, by their inertia they will simply favour job creation *elsewhere*.

## The Welfare Regime

Typically, over the course of this century, as universal welfare provision developed throughout Europe, the self-employed, domestic workers and farm-labourers were the last to be incorporated. While there is still variation within the EU, it is certainly true that, in general, in the northern countries, the social security system is more developed. In Portugal's case, the first significant steps towards the establishment of modern (i.e., universal and contributory) welfare provision in rural areas were taken towards the end of the dictatorship by the modernizing Caetano administration, and only generalized after its overthrow in 1974. As a result, Portugal's population as a whole has been late to enjoy the benefits of universal social welfare provision, and early to bear the cost of its recent restructuring and reform.

The real value of pensions, along with that of many other welfare benefits, practically stagnated for the fifteen years from 1977 to 1992 (Carreira, 1998, p. 71). Today, the welfare regime remains limited and inegalitarian, and still tends to neglect the needs of rural people. Portugal's agricultural labour force is typically dominated by self-employed, often pluriactive small-holders (plus some farm labourers, particularly in the South of the country), and it is precisely these categories that benefit least from the social security system. Distinct rural pensions still exist, though with values broadly on a par with the minimum pension payable to workers who have made no contributions. Low rural incomes and wages limit severely the ability of most farmers to pay their own social security contributions – let alone those of their wives who, incidentally, live longer. This combination of circumstances has left many of the rural older generation with wholly inadequate retirement pensions and a correspondingly miserable standard of living.[17]

Due to the fact that social and health policies are essentially urban-biased, both in terms of expenditure and provision, many rural men and women continue to suffer from a life-shortening or even life-threatening lack of health care. Farming communities typically endure extremely hard living and working conditions, are distant from satisfactory medical facilities, and are almost completely marginal to the existing public transport network. Furthermore, the harshness and loneliness of rural life are aggravated by continued rural exodus: for those who remain in the villages, the growing absence of children, youth and adults of working age, and the tendency towards a withering away – or adaptive restructuring – of social networks and solidarity practices, may contribute to clinical depression or even suicide.

In Portugal's case, since the mid 1980s, welfare transfers, notwithstanding their low absolute level and slow relative growth, have come to play an essential role in the maintenance of basic living conditions among the rural population. Also, many of today's rural residents have spent considerable periods as migrant workers in other European countries, and receive various types of transfers from abroad which both supplement their disposable incomes and may also find their way into supporting such investment initiatives as they or their children may be able to undertake (Portela, 1988).

For many rural households, the availability of conventional health, family, employment and retirement-related benefits are not the only transfers that may bridge the gap between outright penury and spartan survival. If we take into account the discourse, design and deployment of much of the support that state and supra-state agencies provide to farm output and incomes (in the form of investment subsidies, training, etc.) in less favoured rural areas, it would be quite legitimate to see a significant part of the CAP that applies to Mediterranean farming as an extension – inadequate and poorly functioning as it is – to the conventional welfare system. Such an extensive overlap between a formal welfare regime and incentives to agricultural modernization clearly does not characterize the situation in Northern Europe.

## Can Mediterranean Agriculture Evolve from Common Problems to Shared Solutions?

Shared solutions must be based on common problems, and therefore the degree of distinctiveness of the three macro-regions into which Europe is divided – the North, the Mediterranean and the CEECs – needs to be assessed. A distinctive feature of the CEECs is that, regardless of the specificities of the development path each followed, their agricultures remained larger than was the case under European market capitalism. Poorly developed industry in the major cities failed to absorb much of the rural population, and strategies of rural industrialization led to a situation in which 'much of the industrial population lived in the countryside, and, conversely, much of the rural population worked in industry' (Swain, 2000, pp. 1-2). Today, the shares of both agriculture in GDP and food expenditure in family income remain significantly higher than in the EU, while food prices remain relatively low. Despite the substantial changes wrought on agriculture by centralized planning and bureaucratic management, the CEECs did not experience the rural exodus that marked Spain, Portugal, Southern France, Southern Italy and Greece so profoundly from the late 19th to the late 20th century. Central and Eastern European rural areas are nonetheless relatively heterogeneous: over the last 100 years they have been subjected to a diversity of the agricultural strategies, and both smallholder agriculture and a wide variety of distinct farming systems still persist. Nevertheless, there is another CEEC agriculture, dominated by large – often inefficient – farms, struggling to gain a first foothold in international agro-food chains.

That Mediterranean agriculture is itself diverse, has only some similarities with farming in the CEECs, and yet is quite distinct from that practised in the North, does not necessarily mean that all three macro-regions do not share certain concerns and interests. The continuum does suggest, however, that Southern European rural areas are themselves sufficiently homogeneous in socio-economic and cultural terms to be able to jointly identify the 'common ground' on which key aspects of future agricultural and rural strategy may be built. This task needs to be initiated sooner rather than later, for the cultural matrix shared by Mediterranean countries is already under serious threat, as its contours are restructured in the cross-fire of competition between the stronger European economies,[18] by the intense pressures of globalization, and by the energetic promotion and marketing of an increasingly *ersatz* and commoditized version of 'local culture' as the main competitive advantage of rural areas.

The contrasts referred to above suggest the existence of very similar problems as well as common strengths and weaknesses in Mediterranean farming, the more precise identification of which may lead to a joint search for 'shared solutions'. Research is an obvious place to start (Portela, 1994d); however, the main problem in drawing up a Mediterranean rural development research-action agenda would be to know where to stop. A few examples of important agricultural research themes immediately spring to mind: communal property (rights and development initiatives based on such resources), extensive agrarian systems, itinerant grazing, forest development and fire prevention, management of chestnut groves, irrigation systems, farmers' organizations. This agenda could be easily expanded if one were to add broader questions such as local and regional economic diversification, such as those relating to (a) the rehabilitation of rural heritage sites, the promotion of high quality handicrafts, regional cuisine and tourism, and industrial development; (b) the provision of services such as transport, schools, health facilities, including tele-medicine for rural areas; and (c) adult education and professional training, advice in the setting-up of rural micro-enterprises and tele-work.

If, together, the southern countries of the EU are able to seriously reflect together on common farming and local/regional development issues and policies and, above all, if they are able to put any or all of this thinking into practice (even if only gradually and on a small-scale basis), they may discover that they are more alike than they are currently prepared to admit. They may even discover that they have more to gain than to lose if they keep themselves bound by ties of solidarity. However, the shared solutions can be undermined rather easily as Mediterranean power-holders[19] lobby for improvements to EU regional and other development funding, and/or compete with each other to attract private investors, and jockey for position within key market niches both for traditional agricultural output as well as more innovative local products.

Greater supranational solidarity between Southern European members, based on a recognition that their common problems may have shared solutions, and that some of those problems stem from 'common' European policies that disproportionately favour northern interests, could help to transform the rhetoric of economic and social cohesion into a reality. If the huge gap between regions such as Hamburg, Vienna, and Luxembourg, on the one hand, and the Alentejo (Portugal), Epirus (Greece), Extremadura (Spain), and Calabria (Italy), on the other, is to be closed (CCE, 1997, p. 134), neither the magnitude of the challenge, nor the importance of greater intra-Mediterranean solidarity should be underestimated.

Taking the example of agricultural research, it is not enough that common problems be identified, solutions proposed, and research results incorporated into farming practices. Once theory has been transformed into local practice, results have to percolate upwards and influence policy-making, enabling the approach to be legitimized and the potential benefits to be more widely disseminated.[20]

Last but not least, who will ultimately reap the rewards of the value-added generated – locals and/or extra-locals – remains an open question. However, if a Mediterranean rural development policy were to be designed and implemented, the likelihood of local communities and producers being the beneficiaries would certainly increase, remembering that, where population density is low, the creation of even a small number of jobs may have a disproportionately large impact on the local economy and on socio-economic viability (OCDE, 1995, p. 23).

## Four Principles for a Re-engineered CAP

Even before the ink was dry on the 1992 CAP reform, signed under the auspices of the Portuguese Presidency of the Community, a few dissenting voices were already arguing for a 'reform of the reform'. In the intervening years, the criticisms have grown in strength, and today the view that the CAP constitutes a 'common problem', both for the EU and in the wider global context, could almost be considered mainstream thinking. Despite the growing and persistent pressures for reform, it is doubtless still controversial to propose that a shared, Mediterranean strategy might conceivably form part of a future reform of the CAP. While there exist strong forces for change, such as the liberalization of agricultural trade, the enlargement of the EU and the often-public dissent over different aspects of the CAP, they are not necessarily conducive to the type of changes outlined in the present paper. There exists a degree of high-level acceptance that more than marginal adjustments are required: a recent DG VI working group stated openly that:

> the CAP does indeed have to be transformed. It should change from being essentially a centralized commodity policy to becoming a major component of more comprehensive, integrated and decentralized rural policy. [...] The imbalance [affecting the CAP] at the broadest level is its gross over-dependence on the use of market policy at the expense of structural, environmental and rural development measures (CEC, 1996, pp. 5, 35).

The further one questions the continuity of the CAP, its painfully slow evolution and its failure to encompass Mediterranean problems, the stronger the criticisms become. In this regard, Cunha (1997, p. 20), the former Portuguese Minister of Agriculture most associated with promoting the 1992 CAP reform, has stated bluntly that 'a schizophrenic CAP – one that provides direct subsidies to some farmers and denies them to others – is unacceptable' (our translation). Lourenço (1997, p. 16), also a former Portuguese Minister of Agriculture, makes a similar adverse judgement, when he concludes that 'in relation to the Mediterranean areas, the CAP is unequivocally incomplete, marginal and is even the cause of territorial disparities' (our translation). Finally, the recent study mentioned above recognized that for the severest critics, the longevity of the CAP is best explained by:

the power of the lobbies of those who have captured the benefits of the CAP and that it is a sign of the political failures of the EU policy decision institutions and procedures. Such critics view this as a prime example of a democratic deficit in the EU in which a socially illegitimate policy (i.e. one which has indefensible distributional impact and is ineffective in delivering desired objectives) has survived so long. [...Indeed, the ...] CAP has evolved through a delicate political balancing process over thirty years, withstanding major shocks of four enlargements, monetary crises, commodity market crises and budgetary crises (CEC, 1996, p. 4).

In order to redesign the CAP, it is obviously essential to identify precisely who have been its principal beneficiaries and most influential advocates. However, it is also rather important to understand how, historically, its resilience and continuity have been achieved. In this respect the same study sheds further light:

Many of the inconsistencies are the result of years of adding and elaborating policy to deal with successive problems encountered. Rarely have categories of instruments or regulations been removed altogether and replaced by a new measure; almost always new regulations were added to the existing ones. This is nearly always done in a (politically) balanced package of measures in which there are enough decision variables to allow differentiation and [enough] exceptions to allow each member state to 'win' something from the deal. The other occasion on which complications and further scope for inconsistencies arise is at each successive enlargement of the Community. However this political balance has been obtained at the cost of an extremely complicated and increasingly incoherent policy (CEC, 1996, p. 35).

If we juxtapose the imposing array of the CAP inconsistencies, on the one hand, and the prevalent EU discourse emphasizing the key objectives of social equity, balance between regions, regional and local development, environmentally sound practices and high quality of food products, etc., on the other, then what is required is not a 'reform of the reform', but a more explicit and fundamental redesign of the policy. Unfortunately, this goal is too ambitious for the time being and too complex a matter to be satisfactorily addressed here; nevertheless it is important that the shape of such a strategic shift of emphasis be discussed without delay. The rest of this paper is therefore devoted to sketching out the basic contours of such a re-engineered CAP.

In order to achieve the more balanced development of European rural areas implied by a thoroughly re-engineered CAP, two intermediate targets – socio-economic cohesion and a sustainable environment – would have to be met, and this could best be done by operationalizing four guiding principles:

1. Putting people before commodities, and social cohesion before markets;
2. Giving due recognition to multi-functional farming and occupational pluralism;
3. Promoting food quality through locality; and
4. Imposing extensive eco-conditionality.

*People before Commodities, and Social Cohesion before Markets*

A redesigned CAP should – for the first time – embrace Article 39 (particularly point 2) of the 1957 Treaty of Rome, which provides for due consideration to be given to the social structure of farming as well as the structural and natural disparities between regions. It would also need to take very seriously Mansholt's warning, voiced over three decades ago and, apparently, almost immediately forgotten, that 'market and price support policies cannot solve the fundamental difficulties of farming' (EEC, 1968). Thus people, rather than commodities, markets or agro-food enterprises, would form the core of the issue at stake and the point of departure for its solution.

On paper, all types of EU policy are supposed to promote socio-economic cohesion. Given the weight of the CAP in the overall budget, its current policy commitment and material contribution to improvements in the living and working conditions of the least-favoured producers – among whom Mediterranean small farmers loom large – is frankly inadequate. A lower ceiling on the amount of aid per farmer/farm would be a first step in transforming the CAP into a strategy for socio-economic and territorial cohesion. Not only could existing resources be used more efficiently but also the discourse of socio-economic cohesion could begin to take concrete form, if aid were redistributed from farmers already able to resource a competitive strategy, to those that can only adapt if targeted and tapering assistance is provided on a number of fronts.

However, in a redesigned CAP, farmers would not be the only beneficiaries. Rural natural resources can only be properly managed if the men who take the decisions in full- and part-time farming, can call on a wider range of other rural actors for help, notably women, youth and the elderly. Key aspects – both subjective and objective – of rural women's working lives are quite distinct from those of both their urban counterparts and of rural men (CCE, 1994, p. 95), notably their lower pay and higher incidence of unemployment, as well as their shouldering much of the burden of child-raising and caring for elderly relatives. However, this does not mean that women are incapable of contributing further to local sustainable development. On the contrary, they may do so, provided that both current hostility of public administration and the weaknesses of local social organization and management are addressed. Thus, micro-entreprises and handicrafts co-operatives may be implanted or consolidated and generate substantial local impact (Marques and Portela, 1994). Similarly, information dissemination and agricultural training might be much improved (Baptista et al., 2001).

Rural youth, too, face quite distinct problems of potential exclusion (CEC, 1988; Portela et al., 2001) and yet opportunities to more fully integrate youth-oriented policy into the CAP have been consistently wasted, thereby contributing to – rather than mitigating – the problems of youth unemployment and rural-urban migration.

Despite their being seen primarily as contributors to the agricultural crisis, an obstacle to rural development and only 'fit' for so-called 'social farming' (Fragata and Portela, 2000), the elderly could become a much more active resource. Their local knowledge, professional experience, more cautious attitudes towards risk-taking could be deployed in a variety of tasks, ranging from farming consultants and trainers to country park staff, tourist guides, forest guards and fire-watchers – especially if their currently miserable living standard were to be improved. It is worth stressing that the provisions for farmers' early retirement may often constitute a problem more than a solution to revitalizing the agriculture (Baptista et al, 2000).

Rural women, youth and the elderly are all essential to the viability and vitality of rural economy and society. Demographic decline undermines local and regional social cohesion not only by removing active members of the labour force, but also by disrupting and impoverishing key social networks that linking parents, social sector professionals and development agents, for example when a village primary school has to be closed.

Thus, a reformed CAP would have new beneficiaries, namely the poorly endowed farmers and other social categories associated with the rural economy. We have stressed women, youth and the elderly, but the potential of rural newcomers, such as the urban unemployed and returned migrants should also be mobilized (Portela and Nobre, 2001). It would have to find and fund appropriate policy responses to their needs – not as targets of welfare provision or atomized 'policy-recipients', but as members of specific household, village and 'associative' units that would be encouraged to transcend their present client status and become more active contributors to local development.

## Giving due Recognition to Multi-functional Farming and Occupational Pluralism

If agriculture, and the policies to ensure its development, are to be defined from the societal and environmental perspective proposed in this paper, then farming has to be seen as a multi-functional activity. This second principle of a new CAP may provide a basis for a more balanced occupation and use of national and regional space and, via the promotion of such initiatives as rural tourism, rurally-oriented service-providing micro-enterprises, and new productive activities related to forest and water resources, may even contribute to a slowing down of rural out-migration and a consequent mitigation of social pressures in urban centres.

By its very nature, multi-functional farming will always be relatively more labour-intensive, and will continue to depend on close social interaction and co-operation between different local actors. If the current CAP were to recognize this as a key component of Mediterranean farming, policies would have to be reoriented in order to serve a more diverse set of beneficiaries, multi-job holders included. A recent study underlines this challenge by stating that:

> the rate of pluriactivity or multiple job-holding is often very high in the mountains [...Such residents of] the mountain areas studied, are for the most part excluded from Community aid [even though ...] they are often the first to satisfy the multi-functional objectives of agriculture, production, environment, tourism, etc. Recognition of occupational pluralism is very late in coming, and [...] reflects the clash between two logics, one exclusively economic, and the other which seeks to promote a multi-dimensional approach (Euromontana, 1997, p. 4, 20).

From this perspective, a re-engineered CAP would promote the inclusion – not the exclusion – of all the human resources available in a given territory, be it region, micro-region or village. The income and cultural dimensions of this process would have to be addressed in an integrated fashion, on the basis of a clear understanding of how both local social solidarity *and* social differentiation function. Ideally, improvements in the material conditions and organizational confidence of rural residents *should* empower them to make strategic decisions conducive (rather than inimical) to the defence and development of the local cultural resources on which further economic progress and social cohesion in part depend. However, such improvements may also create the conditions for the hijacking of benefits by members of existing or new local elites.[21]

*Promoting Food Quality through Locality*

Having recognized farming as a multi-functional activity, we have to bear in mind that, essentially, it is a process of generating foodstuffs and other useful products that we ultimately either eat or, as in the case of wood, incorporate in some other way into our built and/or living environment. Though modern productivist farming has come a long way, in doing so, it has become detached from its starting point and has shifted its fundamental objective.

Furthermore, the contrast between the techniques and rationale underpinning the mass-production of foodstuffs (on the one hand) and the more 'artisanal' preparation of high quality food items (on the other) is reflected in the greater public concern over the health implications of diet. People today are generally more demanding about the quality of what they consume – in particular what they put in their own and their children's mouths – and this has generated growing demand for clear and unbiased information about foodstuffs, on which people may base their consumption decisions. For these reasons, among others, the principle of food quality assurance, defined as a multi-dimensional concept and practice, embracing aspects such as nutrition, taste, health, aesthetics and biological/regional origin, is an essential objective to be pursued and achieved within the context of a future CAP.

At present, in the act of producing food, farmers may or may not conserve or improve natural resources such as soil, land, water, and landscape, and the same may apply to man-made resources such as historical sites and monuments, traditional rural architecture, terracing and stone walls, hedges and copses. Without financial incentives, combined with specific and explicit policy regulations, farmers have no pressing reason to be more environmentally sensitive or culturally centred: they are left alone to make their own, possibly arbitrary decisions. Farming is, above all, a territorially conditioned activity – what we might term an *art de la localité* (Portela, 1994b) – and this principle should be underlined in a new CAP. If regional space is de-linked from farming, the result might be polluted water, saline soils, eroded slopes, an increasingly homogenous landscape, and empty villages – in other words the death of territory as humanized space.

In response to the demand for authentic, reliable and out-of-the-ordinary products,[22] there has been a proliferation of both territorially specific goods (e.g., raw materials, handicrafts and processed food products, cuisine and beverages) as well as the emergence of a range of services hitherto scarce or unavailable in rural areas. From the shelves of supermarkets and specialist outlets, packaging, labels and logos proclaim the regional, or even more specific local origin of both new and pre-existing items. Undoubtedly, these trends are in evidence throughout the EU and constitute opportunities for the development of the rural areas that a reformulated CAP should not ignore. On the contrary, it should support such initiatives by articulating or, even better, integrating its broader measures with both regional development objectives and regional policy instruments.

*Imposing Extensive Eco-conditionality*

Last but not least, the fourth principle, that of eco-conditionality needs to be applied. From a long-term perspective, it matters not only how much farmers produce but how they cultivate, the quality and variety of the food obtained, as well as the sustainability of the food production system adopted. In this regard, the evidence is overwhelming that more environmentally friendly farming offers the way forward, not only in Europe but also worldwide.

Thus, wherever it may be located, production that is known to cause serious pollution of soil and water, and result in other perverse effects, must be redesigned – a position that has to be explicitly put on the World Trade Organization's (WTO) agenda and vigorously promoted whenever new trade rules, liberalization measures or other such policies are negotiated. More specifically, this emphasis would also require an essentially 'environmental' redesign of the CAP's core – the Common Market Organizations. As pointed out in the above-mentioned Euromontana study, it is necessary to go beyond agri-environmental measures:

> these [...] cannot in fact in themselves modify in any sustainable way practices damaging and harmful to the environment, nor avoid the tendency towards intensification which has become 'spontaneous' after decades of 'farming based on high productivity' [...] It is in the management of the COMs that the fundamental solutions are therefore to be found (Euromontana, 1997, p. 21).

Of course, transferring the principle of extensive eco-conditionality from paper to the fields, pastures and forests will require, among other things, more labour per unit of land, but this is entirely consistent with the aim of job creation in rural areas.

## The Architecture of a Re-engineered CAP: Three Levels of Institutional and Policy Initiatives

Both external and domestic dissatisfaction with the CAP is undeniable. From a Mediterranean perspective, the current CAP is far too distant from mainstream rural society, too focused on productivist farming, too permissive of its own failings, and too bureaucratic to target the least favoured. The current CAP is ultimately and essentially a northern-biased agricultural policy: it is commodity-centred, favouring specific farm products to the detriment of others, and is inherently insensitive to the severe constraints imposed on farming by the harsh physical environment, the risk of commercial loss due to the vagaries of climate and disease, institutional uncertainties, and the structural particularities of rural society. The seriousness of such a charge has been recognized by experts in DGVI, who have concluded that 'the internal legitimacy of the CAP is in grave danger', and that there exists the threat that, 'without further change, the CAP can fracture European construction' (CEC, 1996, p. ix, 2).

The ultimate goal of such a redesigned CAP is clear: a balanced development of rural areas, achieved through greater socio-economic cohesion and environmental sustainability. From a structural viewpoint, such a policy could be envisaged as consisting of a three-tiered process, composed of what we might call 'unionization', 'macro-regionalization' and 'nationalization', each of which would operate – albeit differentially – in line with the four guiding principles (for a re-engineered CAP) presented above. One obvious result of adopting a simultaneous, Europe-wide, yet differentiated and integrated perspective would be to make more transparent the very issues that have often made for difficult negotiations both between current member-states and with candidates for an enlarged EU.

## The Nationalization of the CAP

The issue of the eastward EU enlargement has resulted in the popular belief that its budgetary cost will be enormous and that therefore the 'nationalization of the CAP' is inevitable. So far, it has been assumed that this term essentially refers only to the transfer of a greater proportion of total budgetary costs to member states. It is, in fact, extremely difficult to accurately cost even a first stage of enlargement. Moreover, it is not clear at what phase the key economies of the EU will be in their cycle when the time-table is finalized, how the cake will be distributed, what transition periods candidates will be allowed, or what the effects of all these unknowns will be on the competitiveness of current versus new members, and of an enlarged EU in a global context.

However, such a definition of nationalization presupposes a *quid pro quo*: in return, member states should be seeking not only an extension of their decision-making autonomy within the CAP, where the 'room for manoeuvre' currently enjoyed by Southern European member states is particularly limited, but also the extension of this principle to other key areas, such as regional policy.

Though 'financial' nationalization of the CAP seems quite probable, the scope and extent of the process are far from clear. Were it to transcend its current narrow definition, there would be no necessary contradiction between such a re-engineered CAP and the implementation of the more nationally-focused agricultural policies that are so badly needed. Indeed, it is our belief that the ability of the CAP reforms to deliver their stated objectives of efficient and sustainable agriculture combined with rural development depends crucially on substantial enlargement of national-level decision-making autonomy. Indeed, the re-engineering of the CAP should not be restricted to mere quantitative, i.e. financial or budgetary, questions. For instance, the huge complexity of the CAP – an endless series of interlocking regulations and directives, governing agricultural commodity markets and much more – has to be reduced, too. The 'subsidiarity principle' needs to be explicitly assimilated into a new CAP, with the delegation of as much power and jurisdiction as possible to the national, regional and local levels. Otherwise, neither will the diversity of the local conditions and circumstances be respected, nor will support for the least favoured citizens be achieved.

The decentralization to the national level of much of the focus, formulation and implementation of the CAP would certainly increase national governments' responsibility to their own farm community (in particular) and to rural society (in general). It would be more difficult for member states to argue that Brussels alone constitutes the problem, and it would bring farmers and farmers' organizations into closer direct contact with decision-making centres, providing a more effective forum for resolving specific problems. Above all, perhaps, it would provide a more effective institutional framework in which to remind decision-makers that, despite current neo-liberal policy trends and the ever-growing lobbying power of multi-national corporations, the ultimate and continuing responsibility for the welfare of member states' least favoured citizens – of whom small farmers constitute a significant segment – lies with national governments. For if not with them, then with whom?

*The Macro-regionalization of the CAP*

In this paper we have referred several times to the injustice of a Europe-wide policy in which better-endowed and wealthier farmers, predominantly – though not exclusively – in the North, and particularly those engaged in the production of cereals, oilseeds, protein crops and beef, receive such generous payments from the CAP budget, while the poorer producers, particularly the majority of farmers in the Mediterranean area, receive the crumbs. It is inconceivable that this socially divisive and politically destabilizing feature can be tolerated for much longer, particularly as the eastwards enlargement of the EU approaches.

By recognizing rather than trying to remove the distinctive agricultural features of what we might term macro-regions of the European continent, and by devising policies to realize their specific potential, a new CAP of the type outlined here would contribute significantly to the achievement of a more territorially balanced and socially cohesive agricultural and rural strategy. As a preliminary to identifying the distinct farming types, common contexts and shared problems of such macro-regions, we suggest that the CAP provisions be regionalized into three macro-spaces: Northern Europe, Mediterranean Europe and the CEECs. Of course, no typology is completely water-tight, for there clearly exists a degree of heterogeneity within all these macro-regions, and certain areas would fall on the definitional cusp between categories: for example, the case of the Nordic/Celtic fringe in Northern Europe which has some quite specific characteristics and problems, while sharing others with parts of the Mediterranean and, indeed, the rest of Northern Europe. However, once the European macro-regions have been identified, the corresponding institutional arrangements and policy-formulation procedures, as well as research and development priorities, could be established.

## The Unionization of the CAP

The macro-regional and national institutional and policy dimensions to be built into a re-engineered CAP would necessitate its redefinition at the EU level, too. It is at this level that political discussion, the establishment of general guidelines for integrated rural (essentially, though not exclusively agricultural) development, and the allocation of funds for the macro-regions would take place. Whatever the value of the ideas sketched out here, there remains a demonstrable need for a substantial change in the current CAP, which will require the appropriate structural and financial mechanisms and institutional arrangements.[23] The combination of globalization, the increased volatility of international markets, the continuing ideological and practical imperative of liberalization, and growing concerns over EU governance, all constitute very real and immediate obstacles to the type of re-engineering outlined here. However, without fundamental change, the prospects for southern agriculture remain bleak, indeed. Let us hope that by 2010 we will be able to conclude that the CAP has not merely evolved, but that has been revolutionized in a way that serves, as a first priority, the least favoured in rural society.

## Integrated Agricultural Development as the Starting Point for Mediterranean Rural Development

The main objective set out at the beginning of this paper was to foster discussion on rural development in the Mediterranean context. Yet, paradoxically, or so it may appear, the concept of rural development has not been directly addressed. This was intentional, not least of all because the concept is problematic (see e.g. Newby and Sevilla-Gusmán, 1983; Saraceno, 1994; and Bandarra, 1995). As the CEC (1996, p. 77) concludes 'neither of the words in the term rural development is clear-cut. Both are subject to extensive debate'. On the one hand, rural development is rooted in the urban-rural dichotomy, which today has arguably less relevance than hitherto, while de-ruralization and globalization seem concepts which reflect far more tangible, pronounced and relevant phenomena. On the other hand, rural development can be defined so widely that we may find ourselves either trapped in an academic and semantic maze, or trying to cover so many disparate practical domains that the focus on farming is lost, and farmers themselves are ignored.[24]

Agriculture, too, is a complex concept, and particularly so in the Mediterranean context. Even if seen from an integrated perspective, agricultural development, while not synonymous with rural development, is an essential part of it.[25] We recognize that, in stressing the leading role of agriculture in rural development, we are clearly diverging from the current orthodoxy, in which opportunities are frequently overestimated,[26] recourse to EU-funded niche-market panaceas (such as rural tourism and organic farming) quickly leads to market saturation and the demotivatation of local entrepreneurs, and a triumphalistic discourse of empowerment, partnership and co-operation tends to predominate. However, by bringing farming back to centre stage, we may avoid the even greater risk of seeing agriculture defined once and for all as the exclusive preserve of multinationals and the large and wealthy farmers, with the notion of rural development relegated to a series of residual policies providing a welfare safety net against socio-economic exclusion for pluriactive small-holders and other rural folk.

## Notes

1   An earlier version of this paper was written by José Portela and presented to the Conference entitled 'New Policies for the Development of Countryside in Southern Europe' organized in Athens in November 1998 by the 'Nicos Poulantzas' Foundation and the Institute of Urban and Rural Sociology of the National Centre for Social Research of Greece (EKKE).

2   For the purposes of this article we apply the terms 'Mediterranean' and 'Southern European' rather specifically and interchangeably, to refer to the northern margin of the Mediterranean Sea, plus those regions of the Iberian Peninsula where, climatically speaking, Mediterranean conditions prevail.

3   It should be stressed, however, that there is a tendency for new economic activities in rural areas to be increasingly dissociated from agriculture. Drawing on their income, educational and informational advantages, incoming investors and rural newcomers may be able to 'cherry-pick' the best local opportunities, establish rural enterprises that are not part of the pre-existing mesh of farming and non-farming relationships, thereby intensifying the trend towards the 'detypification' of the countryside to the detriment of the livelihoods of the established population.

4   These candidates (or 'transition economies') for an enlarged EU, whose association agreements are completed, include the following ten countries: Bulgaria, Czech Republic, Estonia, Hungary, Latvia, Lithuania, Poland, Romania, Slovak Republic and Slovenia.

5   Capitalist agriculture will produce anything that will sell and, to this end, has the support of a whole array of other enterprises – particularly in the sphere of farm inputs, commercial distribution and marketing. Here, the term 'productivist' is used to refer generally to capitalist enterprises in agriculture that are oriented almost exclusively towards the attainment of higher levels of farm productivity, but also to point to the narrow definition of productivity typically used (i.e., per hectare, per unit of labour, or per unit of investment), which biases decision-making and priorities in favour of high and simply-measured productivity.

6   For Southern Portugal and Northern Portugal, see respectively e.g. Cutileiro (1972) on the Alentejo region where *latifundia* predominate, and Castro Caldas (2001) on the smallholders and sharecroppers of the Minho.

7   Portugal, along with other Southern European member states, has a tradition of marked out-migration from the rural areas. The persistence of this process may be due, in part, to the absence of policies (a) to staunch the outflow, either in its earlier phases, or more recently, and (b) to build on the experience and potential of returning (e) migrants. About this last point see e.g. Portela and Nobre (2001).

8   The study to which Pinheiro (1997) refers was commissioned by the Joint Nature Conservation Committee and the World Wide Fund for Nature and compared the less

intensive farming systems of nine European countries. The corresponding figures for the remaining countries were as follows: Ireland 35 per cent, Italy 31 per cent, France 25 per cent, Hungary 23 per cent, Poland 14 per cent, and England 11 per cent.

9  A brief summary of these studies is presented in issue N° 5 of the *Newsletter* of the EC's Directorate General VI, October 1998.

10 It should be noted that less favoured status is primarily (though not exclusively) dependent on meeting physical, topographical and/or environmental criteria which are unlikely to be subject to much change.

11 This dependence is best exemplified by the fact that when Southern European companies design and deliver turnkey projects for developing countries, they almost invariably act as 'assemblers' of components provided either by companies from Northern Europe, or from the USA and/or Japan. In this sense, the intermediate status of Southern Europe is reflected in the way it acts as an intermediary between the developed and developing world (Courlet and Laganier, 1986, p. 76). How much its firms learn and gain from this apprenticeship is open to question.

12 Not only do prevailing social relations tend to configure the production, exchange and consumption of knowledge, but also differences in the structures and practice of control, authority and power at the micro-level (between localities), and at the meso-level (between regions which share certain common economic, social, historical and structural characteristics), will be reflected in the extent to which knowledge is commoditized in a given territory, and the values that underpin its generation and dissemination.

13 Such as the creation of *agrupamentos* (groups of producers), for example, among Portuguese fruit-growers, whose production remains in individual hands, but for whom common facilities (such as grading and refrigeration equipment, transport and marketing) are collectively funded and operated.

14 Despite stereotypes to the contrary, farmers do not passively endure unsatisfactory policies. The strategies of their organizations tend to be driven by calls for further aid and for more vigorous lobbying by their regional and/or national governments. This type of action may, in turn, have positive effects both on local cohesion and on the availability of all types of information – technical, procedural, political and organizational.

15 It should be remembered that, wherever new roads are located, they may distribute the benefits of greater accessibility unevenly; thus some town councils in the Interior North of Portugal claim that they are now more marginal than before, and that tourists pass by without stopping, despite the funds spent on restoring historical urban centres.

16 As we have noticed earlier, based on a Portuguese case tudy, the justification for land consolidation projects can be twofold: either the external conditions for agriculture are so bad that reduction of production costs gives a high economic return, or those projects can be used to integrate several functions of rural areas and to improve rural living conditions (Coelho et al,. 1996).

17 If the term 'miserable' appears excessive, it is worth recalling that, in the 2002 electoral campaign in Portugal, the new Socialist Party leader Ferro Rodrigues proposed a radical shake-up (in terms of significantly increased entitlement) of the pension system, if elected. Those couples whose state pensions jointly did not reach the level of the national minimum wage, would have their total income increased: one of them would receive – for the first time ever – the national minimum wage.

18 In the agro-food filière there are numerous cases of the global penetration of the Portuguese market (see Moreira, 2001). A further example highlights the substantial influence exercised by co-operatives and input supply firms from the neighbouring Spanish autonomous region of Galicia over the development of floriculture in North-eastern Portugal (see Gerry et al., 1999).

19  Here, we are not only referring to inter-governmental competition, but also that which pits enterprise against enterprise, farmers' associations against farmers' associations, NGO's against NGO's, etc.; furthermore, firms and other organizations from outside the Mediterranean area may also have strong interests in lobbying in favour of particular uses of EU aid.

20  For initiatives of mutual interest to be designed and for pan-European blueprint policies to be evaluated on something other than a 'parochial' basis, institutional *fora* are also required at the macro-regional level. The EU's existing institutions (or eventually new *fora*) need to focus on both the common traits and peculiarities of the macro-regions suggested below (see the section 'The architecture of a re-engineered CAP: Three levels of institutional and policy initiatives').

21  See Hadjimichalis and Papamichos (1991); Hadjimichalis and Vaiou (1992). Such local economic, political and institutional actors may find it relatively easy to legitimize existing forms of clientelism, adapted to meet the new demands imposed by a changing national and global environment, as the only means of achieving rural development.

22  The growth in demand for what are described as more 'sophisticated' or more highly differentiated products stems from the growth in Northern – in particular middle class – incomes in the 1980s, and the competitive struggle of enterprises (using product differentiation, marketing and other techniques) to create, stimulate, and satisfy such 'niche markets'. Mounting costs and declining margins in industries organized along fordist lines have also contributed both to the development of new commodities and the post-fordist forms of enterprise organization (e.g., flexible specialization, outsourcing and franchising) more appropriate to their production and distribution.

23  In the literature on the CAP reform, much is made of the need for a favourable economic conjuncture. In fact, in contrast to regional and local economic initiatives, for example, the urgent broad structural reforms discussed in the second half of the chapter are longer term and less conjuncturally-driven than many would claim.

24  Despite their continuing numerical superiority in many rural areas, small farmers will become just *one* (albeit marginalized) stakeholder – among many – to be consulted, in a process driven and managed by interests largely antithetical – or at least different - to their own. By way of illustration, in Portugal, this can be observed in the functioning of Rural Development groups in the LEADER programme, in which farmers in general, and small farmers in particular, have often failed to be integrated or to integrate themselves (Afonso 1996; Nogueira 1998).

25  For a discussion of some of the dimensions in which the integration of rural development could be achieved, see Portela (1999a).

26  We should remember that, with regard to SWOT-type diagnoses of rural and local development potential, in a rural society in which social differentiation has far from disappeared, one person's or group's strength may be another's weakness, and what may be an opportunity for some may constitute a threat to others.

**References**

Afonso, F. (1996), *Um Contributo Para a Avaliação do Processo de Execução do Leader-ADRIMAG*, Master's thesis, Universidade de Trás-os-Montes e Alto Douro (UTAD), Vila Real.

Bandarra, N. (1995), 'Specificité du Dévelopment Rural', *Économie Rurale*, no. 225, pp. 15-21.

Baptista, A. and Portela, J. (1995), 'A Pluriactividade como Estratégia de Desenvolvimento nos vales Submontanos de Trás-os-Montes: o Caso da aldeia de Couto de Ervededo', *Estudos Transmontanos*, Ano II, N° 6, Vila Real, pp. 291-333.

Baptista, A., António, P. and Portela, J. (2000), *A Medida de Cessação da Actividade Agrícola em Portugal Continental, 1994-98*, DGDR, Colecção Estudos e Análises N° 10, Lisbon, p. 135.

Baptista, A., Koehnen, T. and Portela, J. (2001), *Estudo de Avaliação das Acções de Formação/Informação para Mulheres Agricultoras e Rurais, Beira Litoral, 1990-2001*, DES-UTAD, Vila Real, p. 85.

Baptista, F. Oliveira. (2001), *Agriculturas e territórios*, Celta Editora, Oeiras.

Caldas, J. Castro (2001), *Terra e Trabalho: Parcerias e Parceiros*, Celta Editora, Oeiras.

Carreira, H. Medina (1998), *O Estado e a Segurança Social*, 'Cadernos do Público' Series, Publication N° 4, *Público* Newspapers, Lisbon.

Coelho, J. Castro, Portela, J. and Pinto, P. Aguiar (1996), 'A Social Approach to Land Consolidation Schemes: a Portuguese Case Study: the Valença Project', *Land Use Policy*, Vol. 13(2), pp. 129-147.

Comissão das Comunidades Europeias (CCE) (1991), *Relatório Sobre a Situação da Agricultura na Comunidade*, Serviço das Publicações Oficiais das Comunidades Europeias, Luxemburg.

Comissão das Comunidades Europeias (CCE) (1993), Comissão das Comunidades Europeias (CCE), *O Apoio às Explorações Agrícolas das Zonas de Montanha e das Zonas Desfavorecidas*, Serviço das Publicações Oficiais das Comunidades Europeias, Luxemburg.

Comissão das Comunidades Europeias (CCE) (1997), *Primeiro Relatório Sobre a Coesão Económica e Social*, Serviço das Publicações Oficiais das Comunidades Europeias, Luxemburg.

Commission of the European Communities (CEC) (1988), *The Future of Rural Society: Report Presented to the Council and European Parliament* [COM(88)371 Final], Luxembourg, July 29th (see also *Bulletin of the European Communities*, Supplement 4/88).

Commission of the European Communities (CEC) (1996), Commission of the European Communities (CEC), *Towards a Common Agricultural and Rural Policy for Europe*, Brussels, Working Group on Integrated Rural Policy, 26th July.

Comunidade Europeia (CE) (1994), *O Papel Económico e a Situação da Mulher nas Zonas Rurais*, Serviço das Publicações Oficiais das Comunidades Europeias, Luxemburg.

Courlet, C. and Laganier, J. (1984), 'Problemas de Desenvolvimento na Europa: o Caso dos Centros Atrasados da Europa do Sul', *Cadernos de Ciências Sociais*, no. 1, pp. 55-85.

Cristóvão, A., Koehnen, T., and Portela, J. (1997), 'Developing and Delivering Extension Programmes', in B.E. Swanson, R.P. Bentz and A.J. Sofranko (eds) *Improving Agricultural Extension – A Reference Manual*, FAO, Rome, pp. 57-65.

Cunha, A. (1997), 'A Reforma da PAC e as Vacas Sagradas', *Expresso* newspaper, Lisbon, 6 December, p. 20.

Cutileiro, J. (1971), *A Portuguese Rural Society*, Clarendon Press, Oxford.

Dries, A. van den and Portela, J. (1994), 'Revitalization of Farmer Managed Irrigation Systems in Trás-os-Montes', in J.D. van der Ploeg and A. Long (eds) *Born from Within – Practice and Perspectives of Endogenous Rural Development*, Van Gorcum, Assen, pp. 71-100.

Euromontana, (1997), *The Integration of Environmental Concerns in Mountain Agriculture*, summary of the study realized for the European Commission Directorate General XI (Environment, Safety and Civil Protection), Brussels.

European Economic Community (EEC) (1968), *Memorandum of the Reform of Agriculture in the EEC*, EEC, Brussels.

Fragata, A. and Portela, J. (2000), 'Agricultores Idosos de Trás-os-Montes: da Exclusão ao Reconhecimento', *Análise Social*, Vol. XXXV (156), Lisbon, pp. 721-737.

Gerry, C., Caldas, J. Vaz, and Koehnen, T. (1999), 'O Boom no Investimento em Estufas na Região de Trás-os-Montes e Alto Douro, 1990-95: O Perfil do "Novo" Empresário Agrícola', *Gestão e Desenvolvimento*, no. 8, Universidade Católica Portuguesa, Viseu, pp. 69-94.

Gusmán, H.N. and Gusmán, E.S. (1983), *Introducción a la Sociología Rural*, Alianza Editorial, Madrid.

Hadjimichalis, C. and Papamichos, N. (1991), 'Local Development in Southern Europe: Myths and Realities', in E. Bergman, G. Maier, and F. Tödtling (eds), *Regions Reconsidered: Economic Networks, Innovation and Local Development in Industrialized Countries*, Mansell, London, pp. 141-164.

Hadjimichalis, C. and Vaiou, C. (1992), 'Intermediate Regions and Forms of Social Reproduction: Three Greek Cases', in G. Garofoli (ed.), *Endogenous Development & Southern Europe*, Avebury, Aldershot, pp. 131-148.

Lourenço, J. (1997), 'Políticas de Desenvolvimento Rural e Políticas Europeias', Documento de Trabalho, Nº 10, Instituto Superior de Agronomia, Lisbon.

Marques, C. and Portela, J. (1994), 'Actividades Tradicionais e Emprego Feminino no Montemuro: Notas de Reflexão Sobre o Desenvolvimento Local', *Gestão e Desenvolvimento* Nº 3, Universidade Católica Portuguesa, Viseu, Portugal, pp. 175-185.

Massot Martí, A. (1998), 'La Agro-Agenda 2000: En Defensa de un Modelo Agrario Europeo?', Volumen Extra Nº 19, XXX Jornadas de Estudio, *Revista de la Asociación Interprofisional para el Desarrollo Agrario* (AIDA), Zaragoza, Spain, May, pp. 14-47.

Moreira, M. Belo (2001), *Globalização e Agricultura: Zonas Rurais Desfavorecidas*, Celta Editores, Oeiras.

Nijkamp, P. (1990), 'Regional Sustainable Development and Natural Resource Use', *World Bank Annual Conference on Development economics*, IBRD, Washington DC, 26-27th April.

Nogueira, F. (1998), *PMEs e Desenvolvimento Local em Meio Rural: o Caso do LEADER I em Portugal*, Master's thesis, Universidade de Trás-os-Montes e Alto Douro (UTAD), Vila Real.

Organization pour le Coopération e Développement Economique (OCDE) (1995), *Créer des Emplois pour le Développement Rural: de Nouvelles Politiques*, OCDE, Paris.

Pinheiro, A. (1997), 'A Agricultura Portuguesa no Contexto da PAC', *Anuário da Economia Portuguesa* (Annual Supplement to the periodical *O Economista*), Lisbon, pp. 194-198.

Ploeg, J.D. van der (1992), 'The Reconstitution of Locality: Technology and Labour in Modern Agriculture', in T. Marsden, P. Lowe, and S. Whatmore (eds), *Labour and Locality*, David Fulton, London, pp. 19-43.

Portela, J. (1988), *Rural Household Strategies of Income Generation: A Study of North-Eastern Portugal, 1900-1987*, Ph.D Thesis, University of Wales, Swansea.

Portela, J. (1994a), 'Agriculture is Primarily What?', in D. Symes and A. J. Jansen (eds), *Agricultural Restructuring and Rural Change in Europe*, Wageningen Agricultural University, Wageningen, pp. 32-48.

Portela, J. (1994b), 'Agriculture: is the Art de la Localité Back? The Role and Function of Indigenous Knowledge in Rural Communities', in B. Dent and M. J. McGregor (eds), *Rural and Farming Systems Analysis – European Perspectives*, CAB International, Wallingford, pp. 269-279.

Portela, J. (1994c), 'Research and Reductionism: Have we Pretty Good Conceptual Tools?' in H. de Haan and J. D. van der Ploeg (ed.) *Endogenous Regional Development in*

*Europe Theory, Method and Practice*, Proceedings of a Seminar held in Vila Real, Portugal, November 4-5, 1991, Organized by European Commission DG VI, Brussels, (EUR 15019 EN), pp. 43-57.

Portela, J. (1994d), 'The Terra Fria Farming System: Elements, Practices and Neglected Research Domains', in Henk de Haan and J. D. van der Ploeg (ed.) *Endogenous Regional Development in Europe Theory, Method and Practice*, Proceedings of a Seminar held in Vila Real, Portugal, November 4-5, 1991, Organized by European Commission DG VI, Brussels (EUR 15019 EN), pp. 227-247.

Portela, J. (1999a), 'A Integração do Desenvolvimento Rural: Pura Retórica?' in C. Cavaco, (ed.), *Desenvolvimento Rural? Desafio e Utopia. Estudos para o Desenvolvimento Regional e Urbano*, Centro de Estudos Geográficos, Publicação N° 50, Universidade de Lisboa, Lisbon, pp. 55-67.

Portela, J. (1999b), 'O Meio Rural em Portugal: Entre o Ontem e o Amanhã', *Trabalhos de Antropologia e Etnologia*, Vol. 39(1-2), pp. 45-65.

Portela, J. (2001), 'Revisiting "Development" in Trás-os-Montes: Between (Neo)Romanticism and Field Observations', paper presented at the VIII National Conference of the Portuguese Association for Regional Development (APDR), Vila Real, 29th of June, 2001.

Portela, J. and Cristóvão, A. (1991), 'Proagri, Extensão e Desenvolvimento Rural: Contributo para uma Reflexão', *Economia e Sociologia* N° 52, Évora, Portugal, pp. 43-74.

Portela, J. and Nobre, S. (2002), 'Entre Pinela e Paris: Emigração, Imigração e Regressos', *Análise Social*, Vol. 161 pp. 1105-1146.

Portela, J., Gerry, C., António, P., Marques C., and Rebelo, V. (2000), *Young People in Santa Marta de Penaguião: from Vocational Dreams to Pragmatism*, chapter 7 of the Interim Report on *Policies and Young People in Rural Development*, Arkleton Centre for Rural Development Research, Aberdeen University.

Saraceno, E. (1994), 'Recent Trends in Rural Development Conceptualization', *Journal of Rural Studies* Vol. 10(4), pp. 321-330.

Swain, N. (2000), 'Post-Socialist Rural Economy and Society in the CEECs: the Socio-Economic Contest for SAPARD and EU enlargement', paper presented at the International Conference on *European rural policy at the crossroads*, June 29th – July 1st, Arkleton Centre for Rural Development Research, University of Aberdeen, Aberdeen.

Chapter 7

# Social Changes and Interest Groups in Rural Spain[1]

Eduardo Moyano-Estrada and Fernando Garrido-Fernández

## Introduction

Today Western societies are undergoing an important process of change characterized not only by globalization and opening of goods and currency markets but also by deep transformations of cultural and political values. Particularly, many of the principles on which government policies were based during the 1960s and 1970s are being revised, and even the viability of the Welfare State itself, as presently constructed, is now in question. Behind the ideological arguments about the greater or lesser role of the state, there appears to be a general consensus on the need for reform in order to reduce budget costs and improve the efficiency of delivery of government policies. It is a matter of political debate whether such reform should be dictated by de-regulation and the retreat of the state and the return of civil society (neoliberal position) or by a reorientation of the regulatory role of the state to guarantee equity and the general interest (social democratic position). Both schools of thought, however, consider reform as necessary.

A policy for agriculture was one of the first to be adopted by the welfare states in Europe during the 1950s, in order to guarantee incomes for farmers and stabilize the agricultural markets. In fact, the Common Agricultural Policy (CAP) was the first common policy to be incorporated into the European Union (EU)[2] after the summit held in 1958 in the Italian city of Stresa. It is not surprising, therefore, that, as a result of the changes that have occurred over the last two decades, the measures adopted by governments to regulate agriculture and the rural society are now out of step with the new problems affecting it. Since 1992, within the EU, reform of the regulatory mechanisms that have defined agricultural policy in the last three decades has been taking place. It is now widely recognized and confirmed in the Agenda 2000[3] that the agricultural policy must be made to function more efficiently, with less bureaucracy and further decentralization. The growth of inequality among farmers and between regions that was not envisaged when the CAP was formulated has to be addressed with new and complementary policies for the increasingly diverse functions of agriculture and rural society (multifunctionality paradigm).

The aim of this paper is to analyze the way in which these changes are perceived within Spanish rural society. To do this we shall take as our starting point the thesis that the perception of change is not homogeneous, but is differentiated according to the different groups that comprise Spanish rural society today. Some, including most farmers and agricultural workers, see it as a traumatic crisis – the end of an era and the loss of their rights and their identities. More forward-looking farmers and people living in rural areas not involved in farming activities see the change as both an opportunity to exploit endogenous resources in different ways and as the beginning of a new era, in which the countryside may be managed in harmony with the plurality of interests coexisting within it.

To begin with, we shall analyze the present process of change, paying particular attention to those factors (economic, social, political and cultural) that affect Spanish agriculture and rural society most directly, and explain new expectations and the emergence of new interest groups. From a sociological perspective, these changes represent a new opportunity structure for the different protagonists that comprise rural society in Spain. This structure offers resources to be exploited by the interest groups according to particular perceptions or interpretations of the process of change. In the second section we shall look at how this opportunity structure is perceived by farmers and non-farmers, and analyze their individual and collective responses to the problems they face.

## The Context of Change

In the case of Spanish agriculture and rural society, the present context of change can be characterized as a series of interrelated factors whose effects are felt at every level, in economic and social life, politics and culture. For the purpose of analysis alone, we shall treat them separately.

### Economic Changes

In economic terms, Spanish agriculture has lost its importance, as witnessed by the gradual decrease in the working population in agriculture (from 13.2 per cent in 1988 to 7.4 per cent in 2000) and the declining significance of agricultural and livestock production in its Gross National Product (GNP) (from 6 per cent in 1985 to 4 per cent in 2000); but it still retains its importance for the dynamism of many rural areas.[4] Much other employment in manufacturing and service industries depends on Spanish agriculture, such as machinery factories and workshops, fertilisers and pesticides businesses, insurance companies, and agro-food industries. What is significant for the argument of this paper is that those involved in activities dependent on agriculture are from an urban industrial background, imbued with a business ethic independent of government subsidies and possessing a non-agrarian view of the value of the countryside. Farmers and non-farmers may share a business relationship, but not necessarily a common system of values when it comes to deciding the fate of the countryside in their local community.

At the same time, the spectacular improvements in telecommunications and roads in Spain have reduced the traditional isolation of rural areas and encouraged the establishment of new industries and services not related to agriculture. Around these activities has emerged a new and increasingly important sector of businessmen and independent professionals with a free market background, whose values are also quite different from those of farmers. Other employments, linked directly to the welfare society, are also creating an unprecedented dynamism in the countryside. This is occurring notably in, on the one hand, the government health, education and social services, and, on the other, in those areas created by the leisure requirements of the population at large, in tourism, second homes, retirement, sport and recreation. The rural population is increasingly involved in such employment, which offers new and non-traditional ways of integrating society and work (Navarro-Yáñez, 1999).

In other words, the Spanish rural society has become more complex, economically and socially, with greater internal differentiation and greater diversity of employment. Landowners' power has been reduced and the growth of new elites encouraged. The new dynamism at the local level is marked by new opportunities for political activism, through co-operation or confrontation between old and new protagonists, depending on their perception of the changes confronting rural society.

## Cultural Changes

Two important cultural changes are evident. On the one hand, there is the rise of so-called 'post-materialist' values (Inglehart, 1977), as ever widening sections of the population are less preoccupied with their material needs and more with quality of life issues, such as deterioration of natural resources; loss of biodiversity; degradation of the countryside; contamination of rivers; and, most recently, food health and quality. There has been an important cultural change in educated public opinion, based on the concept of sustainability minted at the end of the seventies in the now famous Bruntland Report. It lends legitimacy to the demands of new social groups, but, at the same time, introduces important restrictions on the use of the countryside by farmers for agricultural production.

Another change stands out within the cultural context; that of the rediscovery of 'the local', which has been occurring over the last twenty years in Spain at the same time as the rise of globalization. Although apparently contradictory, these processes are, if examined more closely, coherent. The rediscovery of 'the local' is a process of recovering identity, searching for roots and tangible references, of closeness and proximity, in a world that is increasingly globalized and whose physical and social co-ordinates are weakened by being stretched out across the planet. In this context, Spanish people rediscover 'the local', revive the values of their neighbourhoods (pueblos) and seek to remain in them. They attempt to equip their neighborhoods with what they need and exploit the comparative advantages of the advances in technology and telecommunications now offered by the same process of globalization. Local development projects are taking place at what some authors have called the 'interstices of globalization' (Renard, 1999). These projects attempt to value native resources in order to make different forms of development viable and allow rural populations to remain where they are in dynamic communities. This has important economic and political repercussions, and is seen as a revitalizing factor for democracy at the local level (Pérez-Yruela, et al., 2001).

In short, there is a new cultural context in the Spanish rural society characterized on the one hand by a re-evaluating of the countryside, according to criteria that have more to do with the quality of life than with production, and on the other by a revitalization of 'the local' as a central framework of reference for the whole population. Consequently, a new opportunity structure has also been created that is exploited by the various economic and social protagonists, according to their particular interpretation of the changes taking place.

## Political Changes

Some events of the last decade have undeniably affected the framework within which the problems of European rural society in general, and of Mediterranean countries in particular, are addressed.

Firstly, it is worth pointing out the agreements on the liberalization of agricultural markets that took place in the General Agreement on Tariffs and Trade (GATT) (in springtime 1996, in Marrakesh, Morocco) and, subsequently, those signed in the World Trade Organization (WTO) (in autumn 2001, in Doha, Qatar). These agreements have clear political implications; they limit the room for manoeuvre of national governments in maintaining traditional protectionist policies, particularly those affecting agriculture. Some of these implications were already obvious in the reform of the Common Market Organization (CMO) on cereals that was undertaken in the EU in 1992. Implications are evident by the tendency towards a progressive reduction of guaranteed agricultural price and their comparison with market prices, as just established in Agenda 2000. They are also to be seen in the elimination of every type of subsidy for production – in order to reduce agricultural surpluses and avoid their negative effect on international markets – and the implementation of direct payments to farmers.

These political decisions have important economic and cultural repercussions for the agricultural sector. In fact, from an economic point of view, they introduce a new element of competition, confined previously to those sectors that were not sheltered under the umbrella of protectionist policies (mainly, horticulture and fruit). In fact, Spanish farmers, co-operatives and agro-food companies in general are now forced to take competition into account in order to take advantage of the opportunities that wider markets offer. As for the cultural repercussions, these political decisions make important changes in education and training, as well as in the attitude of Spanish farmers towards market and business, necessary.

Secondly, the process of constructing Europe has important political implications for two main reasons. On the one side, the enlargement of the EU toward the former communist countries will represent a high level of expenditure in the EU budget that will require the introduction of important restrictions in the CAP – particularly if this enlargement has to take place without increasing the contributions of the member states to the common budget. On the other side, the process of constructing Europe also implies the incorporation of new policies on the environment, education, research and development, and infrastructure. These policies will have to be funded from the EU budget. The 'agricultural policy community' (Frouws and van Tatenhove, 1993; Daugbjerg, 1997) is now faced with the predicament of having to compete for available resources with other interest groups. These are emerging in a context in which the place of agriculture in European political and social agendas has changed, with the achievement of self-sufficiency in basic food products and the establishment of the principle of open agricultural markets.

Thirdly, the strategic and geopolitical position of the EU in international relations between North and South introduces a very important factor for Spain. The growing tide of immigration from the north of Africa is forcing EU states to change their traditional immigration policies and call for a policy of restricted entry in the short term (viz. the Schengen agreement) and, in the longer term, for an increase in co-operative funding for development in the countries of origin. Such co-operation implies the adoption of measures to open European markets to mainly agricultural and livestock products from these other countries, particularly from countries of the Maghreb, with significant consequences for the Spanish agricultural sector.

Fourthly, an important element of political change and, perhaps, the most far-reaching in its implications in the medium to long term stems from the crisis in the Welfare State that is affecting Western countries and forcing them to review many of the principles that have inspired government policies, including those related to agriculture and rural development. The national budget deficit and, particularly, unemployment and problems related to environmental degradation and food safety, have now to be taken into account in the necessary reformulation of different policies, including the CAP. In the document 'For a necessary change in European agriculture' (Bruges Group, 1998), the Bruges Group pointed out that agriculture policy in future, if it is to have any legitimacy, must take these elements into account. Once self-sufficiency in food is achieved, an agricultural policy which demands government resources to guarantee farm incomes has to derive legitimacy from its contribution to the creation, or at least not the destruction, of employment, equity in the distribution of the CAP direct payments, and the protection of the environment and management of the landscape. These principles that imply a fundamental change in the co-ordinates that have served with reference to farmers and inspired agricultural policies since the fifties are those on which the policies of the future must be based.

In this context, the political debate on the status of agricultural and rural policies for the future is focused in Spain on one important issue: the relationship between rural development policies and policies for regional development. The issue is whether there is any sense in defining autonomous policies for rural development with their own funds and integrating them into an independent institutional framework, or whether it would be more logical for them to form part of more general policies for regional development. Rural areas are tending to lose their special status, and many of the factors on which their development depends, such as road infrastructure; networks for collective ownership of equipment; and state health and education services transcend narrow local boundaries and remain outside the decision making powers of local authorities.

In brief, the above-mentioned context of change, within which debates about the future of rural society and the role of agriculture in its development have to take place, has many elements. These include the waning importance of agriculture in the general economy; the decline of the farming population in rural areas; the diminishing influence of the landed elites at the heart of decision making; the diversification and greater structural complexity of employment in the countryside; and market liberalization. They also include the recuperation of 'the local'; the encouragement of local development initiatives; concerns about the quality of food and the protection of the environment with the achievement of self-sufficiency in food; the restrictions imposed by the process of European construction; and new government policies to overcome the crisis in the Welfare State. A new opportunity structure is created in this context for both individual and collective action by the different social and economic protagonists in the Spanish rural areas. Their actions, however, can be explained, not by any structural determinism, but according to our understanding of how they perceive and interpret these opportunities.

## Different Perceptions of Change in Spanish Rural Society

Perhaps the most important sociological factor in all the changes that are taking place in the Spanish rural society is the increasing complexity of its social structure. The diversification of economic activity, encouraged by programmes for local/rural development; the growth of the agro-food sector and an important tertiary sector; the growing presence of public services linked to the welfare state; the encouragement of the recreational and leisure function of the countryside; and the promotion of its environmental role are important factors in this. Besides the farmers and agricultural workers and their different associations, there are now non-agricultural interest groups with their new dynamism and different perception of the changes taking place.

Even within the agricultural sector itself processes of differentiation can be seen, depending on the type of production and management of different farms, so as on their position in the market and the policies regulating them. The traditional corporatist scenario, based on the principle of common interest among farmers, is being replaced by one of plurality. This is reflected in the diversity of discourses, strategies, and options for organization within the farmers' unions.

There are also important elements of differentiation within the agricultural workers sector, traditionally characterized by strong internal cohesion with respect to land reform. On the one hand, there are those whose integration in the labour market is more or less stable – workers in secure employment and casual workers whose employment is regular, but not continuous – and, on the other, seasonal workers forced into itinerant work with long periods of unemployment. As with farmers, these differences are reflected in the diversity that exists within the agricultural workers' unions.

Therefore, it is clear that the Spanish rural society is today characterized, in sociological and economic terms, by a plurality of interests and by complexity. Nevertheless, when we look at the work of sociologists on social change in rural society we find that the dominant perception is still that of crisis, as though it were a traumatic process for the Spanish rural population.

This approach is, in our opinion, partial and reductive. Most of the Spanish social scientists specializing in rural studies still come from an intellectual tradition with its cultural roots in agriculture. The majority have come from university faculties of agriculture or from the Ministry of Agriculture itself. By choosing to centre their research on the farming sector, they bring to it a perception – generally held by farmers and agricultural workers – of the change in rural society as a crisis because it has deeply modified the framework of reference that guided the economic strategies of farmers and workers, and has breached the social hierarchy in which their status had traditionally been assigned.

Fortunately, for a decade now, a new generation of social scientists from a non-agrarian intellectual tradition attracted by the vitality of some rural areas, has brought a wider approach to studies of rural change. Their research on the rural population as a whole, not just on farmers, provides evidence for the complexity of the social structure, the plurality of interests and diversity in the perceptions of the process of change. While it is true that these social scientists continue to recognize that this change is traumatic for farmers and agricultural workers, they do not extend this view to the rest of rural society in Spain. Their work is interesting, therefore, because in it they reveal different perceptions of social change and show that for some groups it is seen as an opportunity to use the countryside and the landscape in a different way from the traditional one of agriculture.

## Change as a Crisis of Identity for Farmers and Agricultural Workers in Spain

Some sociologists have shown that, in advanced industrial countries, change in agriculture and rural society generally represents a crisis of identity for farmers and agricultural workers, since it questions their whole economic and socio-cultural system in some domains (Hervieu, 1993; Hervieu and Viard, 2001). Besides, one can add that this process of change has broken the traditional unitarian ideal that farmers belong to a homogeneous and strongly cohesive social body with its roots in a common value system.

The present explosion of plurality in the agricultural sector is a reflection of the process of social and economic differentiation that accompanies its full integration into the market, once the old systems of government protection are eliminated. However, there is still a feeling of victimization among certain groups of Spanish farmers, particularly those who have traditionally been in the forefront and who consider it their duty to protect an ideal whose rupture they blame on outside political forces attempting to undermine their internal unity (Moyano, 2000).

In the case of agricultural workers we could add an important rupture in their historical claims for land redistribution and agricultural reform. The changing status of the farming sector and the new symbolic and economic value of land ownership in advanced industrial societies have shifted land reform away from the centre of debate among wide sections of the agricultural proletariat. Many of them are now more concerned with stability of employment – and systems of social protection where these do not exist– and with the improvement of working conditions through collective negotiation.

This situation provokes a general feeling of identity crisis among farmers and workers, even though the way in which the problems are confronted may not be the same. The very diversity and plurality of the responses reflect the reality of a farming social structure that is increasingly differentiated. The different interest groups interpret and respond to the new opportunity structure differently, both individually and collectively.

## *The Individual Responses of Farmers and Agricultural Workers*

There is a diversity of individual responses among the Spanish farmers.[5] Firstly, there are small-time farmers, owners of uncompetitive farms, who have achieved a certain stability through a combination of different sources of income—from the farm itself, the CAP direct payments, employment of family members as workers out of the farming sector, or different forms of state aid, such as unemployment subsidies and welfare payments (for example, retirement payments). For this group, whose role is fundamental for the dynamism and vitality of many rural areas in Spain and other Mediterranean countries and who would be condemned to exclusion if the central European farming model were applied, the present changes offer new opportunities as long as state intervention continues. Without these protective measures, they would find it difficult to survive as farmers. Consequently, it cannot be said that they view change as traumatic because their present position is no worse than it was formerly, when emigration was their main escape.

Secondly, there are farmers with medium and large-sized farms that restrict themselves to following a conservative strategy of collecting the CAP subsidies that, together with income from the sale of their products on the market, guarantee their livelihood at minimum cost and with very little risk. The traumatic nature of the change for this group lies in the growing realization that these subsidies are being cut and may disappear in the future, and that the opening up of the market represents a threat for which they feel unprepared. It is this group, from which until very recently the elite of many Spanish rural communities were recruited, which is affected by the loss of political and economic influence of farming interests. They view the participation of the new emerging groups – particularly the environmentalists – in decisions that affect the fate of the countryside, as interference in the affairs of the farming community. This produces a sort of corporatist retrenchment among them and their discourse becomes that of victimization, a breeding ground for proclamations that demonise politicians. It is, however, a very ambiguous discourse, since it demands state protection at the same time as rejecting any control by official bodies, and defending the right to private ownership of land and the exercise of freedom in farms.

Thirdly, there is an innovative sector of Spanish farmers, drawn from different segments of farming society, which is introducing important changes in its farms. In fact, they are farmers who are opting for new areas of production – the products for bioenergy or the products for the textile or pharmacological industries – and who are exploiting the opportunities offered by new technologies to improve their management. They are also farmers who are developing non-agricultural activities in their farms, such as rural tourism, hunting, forestry, education for children, etc., as complementary sources of income within the framework of the new policies for rural development. Interesting initiatives are to be seen in the sustainable use of natural resources, whether for a more balanced exploitation of agricultural land or a more rational use of chemical inputs to reduce the costs of production. These different responses may be inspired by the principles of a new environmental ethic (Thompson, 1995) based on the recognition of damage done to the environment by intensive agriculture, or by the criteria of 'green' capitalism, concerned about the degradation of natural resources as factors in production. They may simply be, however, pragmatic responses to the new opportunities created by changes in consumer habits, as is happening in the case of the emerging market for organic produce, or by the incentives provided in the agri-environmental programmes of the EU, which offer the possibility of obtaining complementary sources of income (Whitby et al., 1996; Garrido-Fernández, 2000). This innovative sector of Spanish farmers is clearly conscious of the complexity of the changes facing agriculture and of its new position, not yet a leading one, in official agendas. Their attitude in response is not corporatist; there is no retrenchment or looking inwards, but an openness towards the new opportunities offered. Information and training, rather than protection, are sought from the public sector to help them adapt to the new situation, as well as incentives to encourage projects to convert their farms. In their social relationships, they are willing to become involved in joint projects with other interest groups, whether in working with technicians and scientific researchers to study changes in agricultural practices or in collaborating with the authorities in protecting the environment, for example, in fire prevention campaigns.

In the case of the agricultural workers – in 2000 there were 640.200 workers in the Spanish farming sector (almost 50 per cent of agricultural working population), of which more than 50 per cent in Andalucia – their responses are different, but not necessarily exclusive. One response is to stabilize the situation in the labour market by exploiting opportunities arising in different sectors of production and accepting fixed or discontinuous contracts. Another is to take advantage of government systems for social protection that combines unemployment subsidies with programmes aimed at promoting new jobs in rural areas. A third response is to opt for seasonal migration at times of harvest and other agricultural activities.

*The Collective Responses of Co-operatives and Unions*

In the Spanish agricultural sector forms of association play a fundamental role, both in its economic functioning – mainly in the form of co-operatives – and in its providing the central framework of support for union representation, as with the farmers' unions and the agricultural workers' unions.

The strong presence of co-operativism – with its extensive network of co-operatives throughout rural society – and the capacity for mobilization shown by the unions enable both of these types of associations to influence greatly the attitudes and behaviour of farmers and workers; and in doing so, they set themselves up as important intermediaries in the implementation of rural and agricultural policy in Spain. Their political representatives are leaders of public opinion, and their position on agricultural matters resonates in the media and is a point of reference for farmers and workers.

For this reason, it is important to analyse how these associations perceive the present process of change and how they respond collectively, given their influence on the definition of the Spanish workers and farmers' preferences.

*a) Organized Interests in Agricultural Co-operativism*

In Spain, the importance of co-operativism is notable. According to 2000 data, the 3.915 agricultural co-operatives had a total turnover of 1.7 billion pesetas, representing an increase of 30 per cent over 1997. The volume of turnover for agricultural co-operatives represents more than a third of Spanish Final Agricultural Production (calculated at more than 5 billion pesetas for 2000). Its social importance becomes apparent if we consider that in 85 per cent of farms production is linked to a co-operative. Data for 2000 show that 1.1 millions of farmers were members of either one or several co-operatives. Agricultural co-operativism in Spain, however, is characterized by a high degree of fragmentation and internal heterogeneity, with a predominance of small-scale co-operative models whose sphere of activity does not reach beyond the borders of the area where their head offices are located. Nevertheless, in recent years there has been an intense process of concentration through merging or co-ordinating their strategies to enable them to broaden their sphere of economic activity and meet market demands.[6]

According to this situation, it can be said that there is a fairly homogeneous perception of change in the Spanish co-operativism, strengthening the professionalizing tendency in the co-operatives, as well as the introduction of enterprise-oriented criteria in their management to respond to the new climate of competition. The Spanish Federation of Agricultural Co-operatives (Confederación de Co-operativas Agrarias de España, CCAE), the federation covering almost all of the agricultural co-operatives in Spain, clearly shows an entrepreneurial outlook. Co-operatives are companies that have to seek maximum profits in the market. The old mutual ideal is considerably diluted. Its organizational model is increasingly dominated by vertical structures comprised of different farming branches. Unity based on shared economic interests in a given branch or area of production predominates over the old principle of unity based on a sense of belonging to a social movement (see in Table 7.1 the two ideal responses of the Spanish agricultural co-operativism to the present process of change).

**Table 7. 1  Ideal responses of the Spanish agricultural co-operativism**

| | Discourses | |
| **Dimensions** | **Mutualist** | **Entrepreneurial** |
|---|---|---|
| Co-operative Principles | Maintenance of the democratic principle independently of the contribution of each farmer to the volume of productive activity | Allowing farmers more than one vote, depending on their contribution to the total volume of co-operative production |
| | Emphasis on the principle of mutual help and solidarity (to strictly limit the relationship between co-operatives and non-members) | Emphasis on efficiency and the managerial and profit-oriented logic |
| Strategies | Demand for more restrictive legislation to support only those co-operatives that conform to traditional principles | Demand for more flexible legislation to free co-operatives from restrictions imposed by traditional principles |
| | Promotion of the creation of medium size co-operatives integrated into the local area and oriented to the local market | Promotion of the creation of macro co-operatives oriented towards the filiére |
| Organizational models for interest representation | Support for the creation of a multi-sector model based on ideological affinities | Support for the creation of a plurality of sector and branch-oriented models based on economic affinities |
| | Promotion of joint participation of federations of co-operatives with other local groups in new rural development networks | Promotion of interprofessional structures with agro-food companies. |

*Source*:  The table has been elaborated by the chapter's authors.

The new opportunity structure offered by the present process of change is perceived by the CCAE as a challenge to be met with far reaching reform, both legislatively and within the co-operative movement itself. So, these reforms should free the co-operatives of the old clichés and prepare them for competition on the increasingly unavoidable open market to achieve more tangible improvements to the income of farmers. Market logic and competitiveness have managed to replace the old mutualist ideal based on the idea of solidarity at the heart of Spanish co-operativism.[7] As is the case in the rest of Europe (Bager, 1997), the model to be followed now is based on the criteria of organizational bureaucracy, professionalization and a hierarchical division of labour, which are typical of an enterprise-oriented economy that is, nonetheless, socially revamped. Nevertheless, it is already possible to observe in sectors like organic farming and rural development the existence of small co-operatives that stress the importance of the social and mutualist dimension for the local development, and show the rising of tendencies toward internal differentiation within the co-operativism behind the apparent homogeneity shown by the CCAE.[8]

The agricultural co-operatives do not usually express a collective opinion in political debate about future agricultural policy, the modulation of CAP subsidies, or the relationship between agricultural policies and rural development. The heterogeneous nature of their social base leads them to avoid taking up positions that might provoke internal conflict, hence their apolitical nature. The CCAE manages, too, to achieve a difficult balance between its different member co-operatives, and only makes known its opinions as an institution on matters that directly affect the co-operative movement as a whole – for example on the recent reform of the legislation on co-operatives law or on more general matters where there is a consensus. In situations where there is greater dissent, it attempts to adopt an eclectic position and leave the field to the farmers' unions.

*b) Spanish Farmers' Unions*

In contrast to the co-operative movement, plurality is the norm in Spanish farmers' unions. Two ideal types of response may be distinguished here (see Table 7.2). Firstly, there is the 'enterprise response' (agro-food and market-oriented one), espoused by those organizations that mainly represent the interests of medium – and large-scale farms. Agricultural Association-Young Farmers (Asociación Agraria-JóvenesAgricultores, ASAJA) is the best example. They endorse closer integration with the agro-food industry through interprofessional structures in each branch, and a single, rather than multi, sectorial model for the organization of agricultural interests. Farmers are encouraged from ASAJA to adopt new management methods and to continue modernizing production in their farms. However, a detailed analysis of the positions adopted at meetings and conferences by ASAJA reveals a preoccupation with the risks entailed by the exclusive reliance of farmers on subsidies, which are increasingly questioned in the EU and are less secure with the current reform of the CAP.

**Table 7.2  Ideal responses of the Spanish farmers' unions**

|  | **Discourses** | |
| **Dimensions** | **Enterprise response (ASAJA)** | **Neo-peasant response (COAG and UPA)** |
| Conception of farming activity | Market-oriented productive activity | Labour and countryside-oriented activity |
| Status of farmer | Enterpreneur (profesional status) | Farmers with multi-functional status |
| Role of State | Low level of state interventionism (to guarantee the stability of markets) | High level of state interventionism (to guarantee farmer's incomes and correct social and economic inequalities) |
| Function of agricultural policy | Agricultural policy guided by a production-oriented logic | Agricultural policy guided by a non-productive logic and embodied in integral rural development policies. |
|  | Direct payments to farmers to compensate them from the competition of open markets | Direct payments to farmers based on equity. |
| Relationship between agriculture and environment | Environment is perceived as productive resource (green capitalism) | Environment is perceived as important element of countryside dynamism |
|  | Emphasis on the economic dimension of sustainability | Emphasis on the social dimension of sustainability |
|  | The agri-environmental policy is perceived as a complementary income to farmers and an incentive to better use production factors | The agri-environmental policy is perceived as a new source of social legitimacy for both farming activity |
|  | Organic farming is perceived as an interesting market to grow the farmer's incomes | Organic farming is perceived as a way to avoid the social exclusion of small farmers |

*Source*: Moyano et al., 2001.

As for the status of the agricultural policy, they contend that it should remain independent of policies for rural development. ASAJA requires programmes that provide incentives for farmers to modernize more and more their farms and integrate into much wider commercial networks. Future agricultural policy has, therefore, to continue to provide the impetus for modernization to improve competitiveness, particularly in the Mediterranean area, which is backward in comparison with regions in central Europe. That is why ASAJA does not agree with the proposal of integrating agricultural policy into policies for rural development because it would mean subordinating it to a social logic, based on the generation of employment – an impossible objective for modern agriculture, characterized as it is by increased productivity and the reduction of the labour force.

Finally, with regard to environmental policy, the enterprise response is not opposed to it, although it is of secondary importance in its concerns. Problems in the relationship between agriculture and the environment are expressed by ASAJA only in terms of economic sustainability, when the deterioration of natural resources represents a threat to their availability as a factor in agricultural production (in other words, 'green' capitalism, mentioned above).

The second ideal type of response could be described as 'neo-peasant' (countryside-oriented one) because it emphasizes the values of a rural society that has undergone social and cultural renewal, and in which the role of the family farm (a renewed and modern conception of peasantry) should be central and dynamic for the countryside. This response, unlike the first, is voiced by organizations representing the interests of small farmers – Small Farmers' Union (Union de Pequeños Agricultores, UPA) and Federation of Farmers' Unions (Coordinadora de Organizaciones de Agricultores y Ganaderos, COAG), the best examples in Spain. Their policies are concerned not only with production but also with employment and countryside diversification, and they support the horizontal, rather than the vertical, model of representation of agricultural interests encouraging collaboration with other groups in rural society.[9]

Regarding the role of state, they support strong state intervention to regulate market imbalances, and to encourage associations that represent small farmers. There is unanimity among UPA and COAG not only on the usefulness but also on the necessity of applying differential criteria in the distribution of the CAP subsidies. In the face of growing restrictions when assigning resources to regulate the different CMOs, modulation of the CAP subsidies is regarded as necessary because they need to be concentrated on the least competitive farms, if small farmers are not to abandon the agricultural sector. It is also considered useful in restoring legitimacy to agriculture in the eyes of the general population, which views with surprise, if not indignation, the way in which certain groups of farmers amass great fortunes from the CAP subsidies, financed from contributions and given with no undertakings in return and no clear justification. With regard to future agricultural policies, they argue that these should be an integral part of rural development, and include the encouragement of the family farm.[10] For UPA and COAG, criteria should be based not on competitiveness, but on the idea of avoiding the exclusion of small farmers, whose fundamental role in the life of rural areas and countryside, they fully recognize. Environmental policies are viewed, too, as integral to the offering of new opportunities to complement agricultural income, a new way of integrating farmers and countryside into society and a new legitimacy for agricultural policy.

In short, within the Spanish agricultural sector itself there has been an explosion of plurality, reflected in the different responses, individual and collective, of farmers and their organizations to the new problems they face. It is true that the present process of change is perceived as a crisis by the sector as a whole, but responses are diverse, as is to be expected in a social structure that is increasingly differentiated.

### c) Spanish Agricultural Workers' Unions

In relation to agricultural workers' unions, and using as a basis for analysis some characteristic elements of the context of change, such as the loss of the economic and symbolic value of land ownership, the loss of the labourers' identity as a social movement, the improvement of working conditions through collective negotiation and national plans for rural employment, responses can be differentiated into two ideal types in terms of their discourse, claims and strategies (see Table 7.3). Firstly, there is an 'adaptive and reformist response' – represented to a large extent by the unions Agricultural Workers' Commissions (Comisiones Obreras del Campo, CC.OO) and Federation of Land Labourers (Federación de Trabajadores de la Tierra, FTT). Secondly, there is a 'rupturist and radical response' – represented by the Land Labourers' Union (Sindicato de Obreros del Campo, SOC) (Morales, 1997).

The 'reformist response' is characterized by a substantial modification in the traditional position of workers regarding land ownership in order to adapt to the new context of change. It has moved from rejecting any social function of large landowners and demanding official expropriation of land to no longer questioning the principle of land ownership. A distinction is made between large farmers who use available resources appropriately and exercise a recognized social function and those who flagrantly misuse resources and should be penalized.

The 'radical response' continues to question the very basis of the current structure of land ownership, which it denounces as the illegitimate fruit of the historical pillaging of peasant land and, particularly, of the usurpation of their land rights during the disentailments of nineteenth-century in Spain. For the SOC, agrarian reform is still seen as the payment of a historical debt. The illegitimate nature of the current structure of land ownership justifies the use of measures to expropriate large farmers, whether or not they use their resources well.

As far as the social position of agricultural workers is concerned, there are clear differences between the two responses. The reformists accept the gradual conversion of agricultural labourers, and demand that their treatment should be compared with the rest of the working population in improvements in working conditions and salary, and social benefits.[11] The radicals, on the other hand, locate the agricultural workers movement at the centre of the problems of rural society, whose identity and survival require a fundamental critique of the current model of economic development. Their discourse is, therefore, extended to include topics not directly related to agricultural workers, but to the problems of rural society as a whole. Far from following the tendency of reformist unions in assimilating their social bases to those of the rest of the working population, radical unions incorporate other social movements, such as those of unemployed young people, women, and immigrant groups.[12]

**Table 7. 3 Ideal responses of the Spanish agricultural workers' unions**

| Dimensions | Discourses | |
| --- | --- | --- |
| | Reformist and adaptive CC.OO and FTT | Radical and rupturist SOC |
| | They question no longer land ownership | Land ownership is still denounced as the pillaging of peasant land |
| Land ownership | They claim no longer land reform | They continue to claim land reform |
| | They distinguish between large farmers who use resources appropriately, and those who misuse them and who should be penalized | They continue to claim the use of measures to expropriate large farmers, whether or not they use their resources well |
| Identity of agricultural workers | They accept the gradual conversion of agricultural labourers and claim the improvement in working conditions and salary | They locate the agricultural workers movement at the centre of the problems of rural society, and include topics not directly related to them |
| | They assimilate agricultural workers to working population | They incorporate other social groups (unemployed young people and immigrant groups) |
| Collective negotiation | They support negotiations with representatives of agricultural employers to improve the conditions and pay of workers (neocorporatist pact) | They refuse to negotiate with representatives of agricultural employers and use mass action as an instrument of pressure |
| | They are characterized by umbrella organizational structures to participate in neocorporatist agreements | They are characterized by the lack of formality in their organizational structures and they are closer to social movements |
| Government programmes for rural employment | They consider that programmes for rural employment should be directed at agricultural workers exclusively | They demand that such programmes should be extended to the rural population, whether or not linked to the farming sector |

*Source*: This table has been elaborated by the chapter's authors.

With respect to collective negotiation, the reformist unions are characterized by their support for participation in neo-corporatist pacts with representatives of agricultural employers, to improve the conditions and pay of the workers. In this sense, and in accordance with their adaptative ideology, they have accepted the neo-corporatist pact on the understanding that the interests of the workers they represent, who are mainly integrated in the labour market, can be well protected in collective negotiations with employers' organizations.

The radical unions, on the other hand, have refused systematically to participate in these neo-corporatist pacts, because, according to their more radical ideology, the interests of agricultural workers are not to be identified solely with improvements in pay, because there are other problems in rural society. In their opinion, these general interests should be fought for in other ways, in decentralized organizational structures. This explains, for example, the lack of formality in their organizational structures and the systematic use of mass action as an instrument of pressure, as in the numerous marches and occupations of big farms.

Lastly, there is an important difference in relation to government programmes for rural employment and protection against unemployment. While the reformist response accepts that these should be directed at agricultural workers exclusively, the potential beneficiaries having been previously defined precisely from the census, the radical response demands that such programmes should be extended to the whole of the unemployed rural population, whether or not linked to the agricultural sector.

In definitive, plurality is the norm within farmers' unions and agricultural workers' unions. Thus, it is normal that they perceive differently the process of change and answer it differently, too, in order to take advantage of the new opportunity structure.

## Change as New Opportunities for the Non-Agricultural Rural Population

The Spanish non-agricultural rural population has, until now, rarely been asked how it perceives the change affecting agriculture and the countryside. The agricultural background of many social scientists has, as remarked on earlier, influenced their thinking on this change. Since farmers have been the preferred reference group when explaining the change in rural society, it is not surprising that their perceptions have dominated rural research work until very recently.

As we mentioned above, a new generation of social scientists has emerged in the last years, including sociologists, geographers and anthropologists, from different universities and non-agricultural backgrounds, who are beginning to analyse rural change from the perspective of groups unrelated to farming.[13] The work of this new generation of Spanish sociologists is important for this discussion because it demonstrates a perception of change that is very different from that of farmers, and is accompanied not by the traumatic crisis of identity that characterizes their response, but by a noteworthy dynamism. It shows that, for many groups in the non-agricultural population, the present process of change offers great opportunities for revitalizing the countryside and enhancing its use, in accordance with society's new expectations.

The first example of this is in local/rural development programmes, which, channelled through the European schemes (like LEADER), have made possible the emergence of new protagonists in the economic and social life of the Spanish rural communities.[14] New business initiatives, as well as the proliferation of experts and local development agents, and youth and women's associations bring a dynamism to rural areas that helps them to perceive the processes of change differently from farmers. In some instances, these rural development programmes are also making it possible for the more dynamic farmers to be incorporated into development projects, by offering them the opportunity of diversifying their activities and introducing innovations in the management of their farms. Those playing a part in local/rural development, who have hitherto been dispersed and confined to their particular programme, are now beginning to form associations not only to share experiences but also to take action on a wider scale and to participate for the first time as a collective voice at national and international forums where the content of rural development policy is agreed. Because rural areas have, for them, singular features that make them very different from other regions, they are in favour of maintaining policies for rural development independent of those for regional development. They are critical, however, of the agricultural bias of rural development policies, and propose that they should no longer be channelled through government departments of agriculture, but instead should be implemented by interdepartmental, horizontally structured government agencies.

The second example is in the emerging groups under the auspices of the Welfare State, in health, education and social services, which are dynamically redefining the future of rural society and are frequently involved in development programmes. Improvements in communications and the quality of life in rural areas have encouraged these civil servants to live in the villages where they work, and to break the habit, common until very recently, of leaving at the end of the working day. Their importance for education and health policies and for the location of schools and health centres cannot be denied; and their role for the future of rural areas is very often much greater than that of rural development. That is why these new social actors support more regional rather than independent rural development policies to solve the problems of rural areas.

Lastly, those from urban areas, such as holidaymakers, weekend and day visitors, ramblers and those interested in country sports, are encouraged by the attention paid to the leisure and recreational functions of the countryside (García-Sanz, 1999). While rediscovering the old rural folk traditions, they are, at the same time, managing to introduce typically urban patterns of cultural behaviour (night life, mass car use, etc). The involvement of young people, too, in the new types of agriculture – in, for example, organic farming, where they often work with local environmentalists – brings a non-productionist agricultural slant to the exploitation of natural resources. This distances them from the traditional discourse of farmers and provokes internal divisions in the local agricultural sector.

The present context of change offers opportunities for all of these groups to revitalize rural society, allowing them, by their participation in local policy-making, to influence decisions taken at the local level. It is increasingly common to find councillors in rural authorities who are doctors, teachers, social workers, representatives of environmental groups, etc., taking their place alongside professionals and businessmen outside agriculture in the new local elites.

The responses of these groups are, however, varied, and this demonstrates a diversity of interests that must continue to be an object of research for social scientists, to enlarge our understanding of the social and economic dynamics of rural society today in Spain and the southern countries. Social scientists have an interesting laboratory in which to discover whether or not a new rural identity is emerging, no longer characterized exclusively as agricultural, but rather as a synthesis of different activities and professions, including agriculture. These come together as development takes place in small – and medium-sized centres of population, and have a special connection with the area. The question to be resolved is whether this identification between the different groups in rural society is strong enough to allow one to speak of the existence of a new rural identity. What may exist, on the contrary, are groups with distinct, and unconnected, identities, with no-feeling of belonging to a cultural community and no interests in common.

## Conclusions

An analysis of the processes of change in rural society can contribute to a better understanding of social dynamics in contexts of new opportunity structures. These are exploited differently, according to the particular perception and interpretation of the process of change, and strategies developed, individually and collectively, to confront the problems that arise. Several conclusions can be derived from the analysis offered in this paper.

In the first place, in the contemporary context of globalization and the reform of traditional protectionist policies, Spanish rural society is no longer a world apart. It is increasingly influenced by the wider society, reproducing the dynamism and diversity of interests characteristic of open societies.

Secondly, it is understandable that Spanish farmers see the process of change as a crisis of identity, since it is producing a radical transformation of the frame of reference within which they have acted in the last four decades. The gradual integration, however, of agriculture into the market, once the umbrella of protectionist policies has been removed, produces economic and social differentiation among farmers themselves. This is reflected in the different perceptions they have of the context of change and their different responses to it. The principle of unity in the agricultural sector, more symbolic than real, has given way to that of plurality, as shown by the diversity that exists within the Spanish farmers' unions.

Thirdly, Spanish agricultural workers are undergoing a profound modification in their reference system both socially and symbolically, owing to the changing status of agriculture as an industry and of land ownership as an economic resource, and to the reform of the labour market. In this context, the dominant response is to adapt to change and exploit the new opportunities it offers to improve the conditions for agricultural workers. In certain Spanish rural areas, however, responses are more radical, the demand for land reform is maintained and a new identity for workers sought, within the framework of the new values emerging in the countryside.

Fourthly, groups not directly related to agriculture are establishing themselves alongside farmers, and are bringing both a new cultural dynamism and values that are different from those that have dominated Spanish rural society until now. The social structure has become more complex and relationships between the different groups more dynamic, sometimes through co-operation, and other times through conflict over the definition of the countryside in Spain.

Fifthly, in a socio-cultural context characterized by self-sufficiency in food, the growing popularity of post-materialist values, the demand for a sustainable development model, the reaffirmation of the local in the face of globalization and the need to redirect the role of the welfare state, the countryside is defined as multi-functional. This has important repercussions for the principles that have inspired official policies and particularly agricultural policy, whose justification has been to exploit natural resources to provide food. In the new context, agricultural policy has to seek new legitimacy for farmers to continue to receive subsidies. In the debates on the future of agricultural policy and rural development, some emerging elements are the generation, or at least not the destruction, of employment, equity in the distribution of payments, and food quality, its contribution to the management of the land and the protection of the environment.

In short, in a context marked by a diversity of expectations and plurality of interests in rural society, policies, too, should be different. Old agricultural policies directed to market regulation and structural modernization are being reformulated according to the paradigm of sustainability and ecological modernization, and new policies are emerging to regulate the many functions of the countryside. Old and new policies and protagonists coexist in this period of transition and give the Spanish rural society a dynamism, unknown before, that offers a new opportunity structure to the different social groups. For the population of rural areas, individually or collectively, this structure merely provides a stage. Its proper subjects are those who, through their particular ways of interpreting it, define their preferences and exploit the resources at their disposal.

## Notes

1    Jan Treacher has translated this text into English language.
2    Although the European institutional construct has been named European Economic Community, European Community and European Union in different periods, we will always refer to it in this article as the European Union.
3    Agenda 2000 was passed by the European Council in Berlin Summit, June 1999.
4    In some rural areas of the southern regions of Spain, more than 40 per cent of the

population is working in the agricultural sector.

5     In 1998, there was 1.21 millions of farms in Spain, and 2.54 millions of people working in agriculture (farmers and workers) (926.000 UTA per year).

6     See in Moyano et al., 2001 a wide analysis of the Spanish agricultural co-operativism.

7     The CCAE has just supported the new Act of Co-operatives (1999) in which enterprise-oriented logic is the dominant principle.

8     On new ways of economic co-operation, see Sacco dos Anjos and Moyano, 2001. In this article, the authors analyse the *condominios* as alternative to large co-operatives in Santa Catarina (Brazil).

9     Examples of this are the Rural Platform set up by COAG, and the collaborative agreement between UPA and the environmental association for the protection of birds, SEO (*Sociedad Española de Ornitología*).

10    The 1999 Agricultural Yearbook edited by the UPA was dedicated to the topic of 'Redefining Family Agriculture'.

11    In 1997, CC.OO and FTT-UGT signed a social pact with the conservative Government – named *Acuerdo sobre el Empleo y la Protección Social Agraria* (AEPSA) – to reform Spanish legislation on agricultural workers in order to extend to them the general rights of workers. This pact was not signed by the SOC. In 2000, the reformist unions (CC.OO and FTT-UGT) participated in the creation of an agro-food workers' union.

12    Recently, the SOC has changed its name, adding the words '*y de los trabajadores rurales*' (and the rural workers). Furthermore, it allows, now, migrant labour to join it.

13    At the last three meetings organized by the Spanish Federation of Sociology (FES) in 1995, 1998 and 2001, more than two-thirds of the papers presented in the rural sociology workshop were related to local development, the environment, rural tourism, crafts, fishing, mining, the management of natural parks, etc., all of which, while not excluding farmers, are new topics unrelated to agriculture.

14    The European Commission Initiative LEADER and the Spanish operative programme for rural development (PRODER) are programmes of bottom-up development that extend throughout the country a network of 233 Rural Development Groups in areas with fewer than 100.000 inhabitants. For example, in Andalucia, that is the biggest region of Spain, 666 of the 770 municipalities (85.4 per cent of the total number, 88.6 per cent of the land area and 41.8 per cent of the population) were affected by rural development activities under these two programmes, during the period 1994-1999 (see Garrido-Fernández, Mauleón y Moyano-Estrada, 2002).

## References

Bager, T. (1996), *Organizations in Sectors. Explaining the Dissemination of Formal Organization in Economic Sectors*, South Jutland University Centre, Esbjerg.

Bruges Group (1998), *For a Necessary Change in European Agriculture*, MAPA, Madrid.

Daugbjerg, C. (1997), 'Reforming the CAP: The Roles of Policy Networks and Broader Institutional Structures', *ALF Working Paper*, South Jutland University Centre, Esbjerg.

Frouws, J. and Van Tatenhove, J. (1993), 'Agriculture, Environment and the State: the Development of Agri-environmental Policy-making in the Netherlands,' *Sociologia Ruralis*, Vol. 32(2), pp. 220-239.

García-Sanz, B. (1999), *La Sociedad Rural Española ante el Siglo XXI*, MAPA, Madrid.

Garrido-Fernández, F. (2000), *La Cuestión Ambiental en la Agricultura*, Unicaja, Málaga.

Garrido-Fernández, F. et al. (2002), 'Rural Spain', in K. Halfacree, I. Kovách and R. Woodward (eds), *Leadership and Local Power in European Rural Development*, Ashgate, Aldershot, pp. 173-202.

Hervieu, B. (1994), *Les Champs du Futur*, Boulin, Paris.

Hervieu, B. and Viard, J. (2001), *L'Archipel Paysan*, L'Aube, Paris.

Inglehart, R. (1977), *The Silent Revolution: Changing Values and Political Styles among Western Publics*, Princeton University Press, Princeton.

Morales, R. (1997), 'Desarrollo y Transformaciones Históricas del Sindicato de Obreros del Campo (SOC) (1976-1994)', *Sociología del Trabajo*, no. 32, Siglo XXI, pp. 31-50.

Moyano-Estrada, E. (2000), 'Farmers' Unions and the Restructuring of Agriculture', in W. Grant and J.F.T. Keeler (eds), *Agricultural Policies*, Edward Elgar Publishing, Northampton, pp. 217-234.

Moyano-Estrada, E. et al. (2001), 'Federations of Co-operatives and Interest Organized in Agriculture. An Analysis of the Spanish Experience', *Sociologia Ruralis*, Vol. 41(2), pp. 237-253.

Navarro-Yáñez, C. (1999), 'Women and Social Mobility in Rural Spain', *Sociologia Ruralis*, Vol. 39(2), pp. 222-235.

Pérez Yruela, M. et al. (2001), *Las Nuevas Concepciones del Desarrollo Rural. Estudio de Casos*, CSIC, Córdoba.

Renard, M.C. (1999), *Los Intersticios de la Globalización*, Chapingo University, Chapingo.

Sacco dos Anjos, F. and Moyano, E. (2001), 'New Ways of Economic Co-operation. The Case of Condominios in Santa Catarina, Brazil', *Journal of Rural Co-operation*, Vol. 29(1), pp. 25-45.

Thompson, P.B. (1995), *The Spirit of the Soil. Agriculture and Environmental Ethics*, Routledge, London.

Whitby, M. (ed.) (1996), *The European Environment and CAP Reform*, CAB International, Wallingford.

Chapter 8

# Local Democracy Put to the Test of Negotiated Development

## Introduction

One of the most remarkable aspects of the new development policies is the obligation placed upon the social actors to take concerted action for their implementation. We find a similar obligation in a great many European policies applied in the past ten years to agriculture (Regulation 2078/92 governing agri-environmental measures in 1992) or to rural areas (the Habitats Directive, also in 1992). This trend offers a concrete illustration of the 'principle of subsidiarity' (Millon-Delsol, 1993) or, more generally, points to the emergence of a new form of 'governance' tending to involve different parties from the public and private sectors faced with increasingly complex problems in the decision-making process. In every case, the determination to make consultation[1] an integral part of the decision-making process is manifest and, irrespective of the position of the actors in this process, the search for a compromise or agreement goes far beyond mere procedural requirements.

This new facet acquired by public policies favours the emergence of local arenas derived from these negotiations and consultation procedures. The French agri-environmental experience, based on the ability of the different actors in rural areas to work collectively in debating questions related to the treatment of nature, offers a perfect illustration of the possible outcomes and forms that may be produced by development increasingly founded on negotiation. The aim of these measures was to integrate environmental preoccupations into the modes of agricultural production. But, to do this, it was necessary to mobilize expert opinion to provide a legitimate basis for public policy. And this is where the result – negotiation – coincides with the form, because the fact of contributing to a decision-making process supposes having recourse to expert opinion, conflicting or not, in order to develop a point of view. This recourse in turn raises the question of the relationship between a 'technical democracy',[2] emerging through this 'negotiated development', and the traditional local democracy of rural societies of which the local politician is the principal representative. The relationship with the land, the fact of belonging to a territorial community underpin the very legitimacy of this mode of representation. We may wonder whether the 'democratic' expansion triggered by the procedures of 'negotiated development' can strengthen or, on the contrary, weaken traditional local democracy. Indeed, the scope for the exercise of this new type of development is rarely limited to a communal area and the principles and forms of action present particular characteristics illustrated, for example, by the use of scientific and technical expertise.

To illustrate this proposition, we shall first examine the context which encourages such a transformation in the forms of local governance: without being the direct consequence, this change owes a great deal to the emergence of the notion of 'sustainability' or, at the very least, it cannot be understood without taking this new context into consideration. The French agri-environmental experience will then allow us to illustrate through a case study the outward signs and possible outcomes of this type of development. And, lastly, we shall attempt to see the consequences, for the farmers but also for rural societies as a whole, of the emergence of such 'local arenas'[3] built on negotiation procedures.

### Territory, Development and Agriculture: the New Context of 'Sustainability'

For the past ten years or so, it can be said that a 'new situation' for agriculture has been emerging, driven by a large number of heterogeneous factors that tend to revolve around the notion of 'sustainability'. The idea of sustainable agriculture is presented as a unifying notion that expresses both an observation (something in agriculture is changing) and, simultaneously, a prescriptive attitude (agricultural activity must change). The consensus which seems to be forming around this idea is tending a little quickly to suppress memory of a time when the relationship between agriculture and environmental issues was, in France at least, the subject of a veritable political taboo and, as such, was banned from the official agenda. This new consensus can lead to an excessively conventional discourse and make it difficult to identify and analyze what is really at stake in the use and implementation of this notion, but it is also well suited to calling established positions into question. The notion of 'sustainable agriculture' is also a subject of both theoretical and practical investment, which varies depending on the social actors claiming to take their inspiration from it, the sectors of activity where it is supposed to apply, and the countries in question (if we consider North-South relations, at least). More than a concept offering immediate guidance for action and decision (the search for indicators tries to satisfy that objective), the notion of sustainability primarily reveals the crisis of the notion of post-war development and the emergence of a new, multi-faceted demand of an ecological, economic, social and even ethical nature.

Whether related more generally to the question of 'development' or more specifically to farming, the notion of sustainability is used by the scientific community and by the media, and even finds an echo among the general public: this extremely wide use, fraught with multiple meanings, must be seen as the expression of particular collective expectations. Thus, in countries enjoying high levels of technology, it is the consumers who push for the adoption of new compromises, either with the producers themselves or with the public authorities,[4] a trend given a significant boost by the more recent crises (mad cow disease, dioxin-contaminated chicken, etc.) affecting European agriculture.

This notion seems to enjoy a broad consensus and finds expression in a phenomenon shared by all developed countries, namely: the growing trend to include environmental issues in the drafting of public policy. More than any other industrial sector, agriculture has long been subject to this type of policy for reasons of food safety or the conquest of new markets. It does not, therefore, escape an evolution characterized in all places by two aspects: firstly, a mobilization of interdisciplinary scientific knowledge (ranging from agronomic to genetics, from biology to ecology) and, secondly, an expression of consumer sensibility (ranging from product quality, and its consequences on health, to aesthetic considerations related to nature and the countryside). Concretely, this process began in Europe in June 1992 with the Agri-environmental measures which placed environmental preoccupations (a movement launched ten years earlier in the United States with the 'Farm Act' in 1980) at the very heart of the Common Agriculture Policy (CAP). All these policies, which to a greater or lesser degree claim to be founded on the notion of sustainability, form part of a context disturbed by the globalization of the economy. As such, the problems are not new but it is becoming urgent to find a solution: structural overproduction; increasing burden of social, ecological and, above all, budgetary costs; saturated, chaotic, increasingly less regulated markets. The official position of the Organization for Economic Co-operation and Development (OECD) countries is relatively clear and shared by its members: the choice of sustainable agriculture supposes a reconciliation between the economy and the environment (OECD, 1994), i.e. a renewal of the social regulation of this sector through market liberalization (end to subsidies, alignment on international prices with a view to putting a stop of overproduction) and the promotion of high-tech agriculture.[5]

If we had to summarize and, undoubtedly, oversimplify the debate about the question of sustainability applied to agriculture, we could ask the following question: is there too much, or not enough, market to manage natural resources? Globally, we can say that the paradigm of sustainable agriculture, i.e. the socio-political project underlying it, calls for a new phase of agricultural modernization. But does this modernization bear the seeds of a new model of production or does it correspond to a particular stage in the selective modernization that began more than 30 years ago in Europe?[6]

Therefore, above and beyond a consensus that raises more questions than it answers, there exist ambiguities and even controversies about the impact and tools of regulation, depending on the place granted to market forces. It is even possible to talk of uncertainties about the risks themselves[7] due to a confrontation between environmental and agricultural policies in a market economy. In other words (in order to summarize and simplify once again), are we faced with a simple correction or even a new adjustment of a production model in crisis (the search for 'more rational fertilization' or 'integrated production') or, alternatively, are we dealing with a process leading to a radical break with an outdated model of production?

In this sense, the context, whose very novelty is denounced by the notion of 'sustainability', is particularly well suited to the idea of negotiation to the extent that the prescriptive conception, which then underpins development, is immediately put to the test of objective fact and action. And this test is all the more necessary as this reference to sustainable development is not sufficient to clarify the ambiguities or, in any case, the ambivalence which can be identified, for example, with respect to sustainable agriculture, namely:

- faith in technological innovation, but also an environmental diagnosis which supposes a critical examination of the notion of 'technical progress' applied to development;
- the dogma of market liberalization, but also a tendency to impose voluntarist industrial price policies (fiscal measures, taxation, etc.) to oblige the different parties to adopt a more rational form of behaviour that takes account of the environmental dimension;
- an economic optimization by the market, but, at the same time, the greater use of a system of obligations between actors ('direct aids') which anticipates a management of natural resources pursued independently of any market regulation;
- the inclusion of local particularities down to individual parcels of land, but also considerations which acquire form and meaning at a planetary level ('Global Change');
- a process of consultations and negotiation within extremely localized frameworks, but also an extension of standards (of the 'Nitrates Directive' type) and exaggeration in the definition of territories.

The French agri-environmental experience will soon be celebrating its tenth anniversary. We are entitled to see it as a kind of experiment in the application of 'sustainability' to the farming sector to the extent it has set itself the goal – in its justification, at least – of making the demands of the economy (the fight against overproduction) more compatible with environmental requirements. It offers a perfect illustration of how dialogue and compromises are vital for drawing up standards that are both shared and efficient for collective action. As such, it makes it possible to raise the question of the forms of democracy required by this type of development and illustrates the stakes revealed by their experimentation in contemporary rural societies.

## 'Negotiated Development': Forms and Possible Outcomes

The fact which has interested most observers of the French agri-environmental experience is the creation of discussion forums and negotiations related to the definition, at a local level, of the terms and conditions of action. This experience made it necessary to set up mechanisms leading to a confrontation between the representatives of the farming industry (both the farmers and the administration) and the other users of nature (hunters, fishermen or landowners and actors in local public life such as local politicians or associations related to the ecology movement). The use of contracts, agreements, partnerships, charters – which political scientists call 'Contractual Public Activities' (CPA) – is nothing new to industrial societies nor something new to French agriculture. The system of agricultural co-management is related, to a certain extent, to this type of agreement. Similarly, the development of rural areas has, for more than twenty years, given rise to such forums bringing together a number of actors around questions of economic development (the Rural Development Plans and Country Contracts) or related to the protection of nature (regional nature reserves). In other words, the confrontation between the representatives of agriculture and the other stakeholders in rural areas constitutes a founding principle of local democracy in the French rural environment. And yet, a feeling of social innovation is emerging from the agri-environmental experience which derives, it would seem, from two facts: firstly, the fact that the materiality of nature (in this case, the physical and ecological dimension of farming) has become a central aspect of social interactions: the social actors defending environmental protection assume to play an unprecedented role in this forum; secondly, the fact that these negotiations use expertise as a basis for the justification of public intervention. In this case, the representatives of agriculture become one category of experts, among others.

## For Agriculture, an Innovation

The novelty resides, first of all, in the radical nature of the process: never before have production practices been subject to appraisal so dependent on a consultation process where the representatives of the farming industry become experts among others; never before has the treatment of nature been so exclusively and so generally the subject of contractual limits imposed on the practices of the farmer. These forums, whose membership varies with local situations, have invented new intervention mechanisms offering an alternative to those previously prevailing around market logic or government directives. The agri-environmental procedures, through the discussion of contractual terms and conditions, of the exact scope of application, have opened agriculture up to the scrutiny of a wide variety of rural actors and offered the possibility of experimenting with new forms of compromise and even new systems of action.

In other words, agriculture must no longer merely give account of how it occupies the land but also, and above all, explain how it makes good use of it. The idea of a collective right to scrutinize the way private space is used – in this case, individual practices on parcels of land – has acquired certain legitimacy in the agri-environmental experience. But it is common knowledge that control, based on systems of appraisal, forms an integral part of sustainable agriculture projects to the extent it associates collective arbitration with practices of the most private nature (Billaud, 1995).

The development model underlying the agri-environmental experience also carries the seeds of a radical new departure in the world of farming from another point of view: productive practices are no longer founded on a technical script based on laboratory experiments or a regulatory framework but on a social consensus derived from heterogeneous locally-produced knowledge (agronomic, certainly, but also ecology and, within it, ornithology, phytosociology, etc.). These new joint environmental and agricultural management systems create a wide range of situations, all the more so as the agri-environmental procedures are particularly flexible at a prescriptive level and consequently widen the range of possibilities in the modes of social construction of nature. The systematic use of negotiation procedures to determine the objectives of action, the use of inter-actor obligation systems (individual contracts) to implement them illustrate one of the answers developed by modern societies to the question of the social treatment of nature. The confrontation between rural social groups and agri-environmental procedures has made the physical and ecological dimension of agriculture a central issue in social relations.

The standards of productive efficiency, which used to be more a question for agronomy and economics, must now co-exist or compromise with other principles of action and management, such as the need to take account of the specific nature of the environments to be maintained, to preserve outstanding habitats or landscapes, implying a confrontation process with other scientific communities liable to call into question the habitual reference system of farming practices. This constitutes a form of submission on the part of agriculture to other legitimate claims (of hunters, fishermen, ecologists, urban dwellers, etc.) and, as such, the end of a certain hegemony enjoyed by this industry in rural development. We can also see a sort of reactivation, or even renaissance, of the rural environment to the extent that negotiating always corresponds to a collective apprenticeship, i.e. a closer contact with the 'other'. By reflecting a certain sociological diversity, much more faithful to contemporary rural society than what prevails in traditional sector-based policies, the agri-environmental action systems may provide an opportunity for the local reappropriation of environmental issues to the extent they have been, through these forums, socialized and, perhaps, humanized. If there have been changes, they affect all the partners, i.e. the supporters of agriculture as well as the advocates of environmental protection.

*But, it is True, of Limited Impact*

This process of 'negotiated development' naturally takes more account of a project or dynamic current than provides a description of what more generally happens in the field: the situation described in these pages usually produces disappointing results. In other words, the 'natural' point of view supposed to guide public action and destined, in this respect, to become a 'central figure in the common good' (Boltanski-Thévenot, 1991) has rarely occupied such a position. Let us take the case of local agri-environmental operations. They can be broken down into three key moments, each offering a possibility to make the environment and the treatment of nature a justification for collective action: the definition of a scope of action, the definition of contractual terms and conditions, the choice of farmers able to enter into a contract. Logically, the choice of the contracting farmers depends upon the definition of the scope of action and the terms and conditions to be applied.

The scope and conditions must be drawn up according to the framework provided for under European regulations, on the basis of a set of ecological criteria initially determined by an expert appraisal of the situation. In fact, this virtuous framework is rarely verified. Most frequently, the scope is determined on the basis of the financial resources, i.e. to make demand correspond to supply as closely as possible. The specifications are drawn up with a view, not to selection based on biological excellence, but to ensuring the equality of all the farmers before a subsidy system. Thus, the choice of contracting farmers is merely based on social criteria in which the 'natural' point of view is irrelevant; the farmer is chosen not for his environmentally-friendly practices but by virtue of his being the principal operator or, in cases of strong demand, a resident of the application zone, etc. Compared to the results of changes in practices or technical innovations, the agri-environmental experience could seem vain, just as it was by no means surprising that the ecology associations should hesitate between a protest-based culture and opting for active involvement.

## For Rural Societies, a Dynamism that is Both Unprecedented

And yet, the large echo given to these initiatives – both in the mass media and in the scientific community – is due to the fact that, above and beyond the weaknesses or deceptions, there emerged from these programmes a collective experience of an unprecedented dynamism in the construction of a social relationship to nature. The individuals involved in these negotiations about the definition of specifications were confronted with the need to bring together scattered information, to translate it into a form suitable for discussion, to compare farming practices and environmental qualities. In a word, they built socio-technical 'assemblies'[8] giving official status to social and scientific uncertainty and preaching constant adjustment and compromise as a guide to public action.

If the question of nature has most frequently been repressed in these contractual procedures based on the principle of negotiation between a large number of partners, it nevertheless remains true that the appropriateness of its inclusion is now at the heart of the rural question. Agriculture can no longer treat the question of its relationship with nature implicitly; by accepting it as a specific problem, agriculture is not only obliged to think about its own practices but also to build with the other stakeholders (and users) of rural space a new way of 'living together'.

Because, behind all this, what is at issue is the question of economic development and its inclusion in rural territories and, more particularly, the functions to be carried out by the farming industry in these territories. The debate is sufficiently important for the elected representatives of local communities to have been particularly involved in the negotiations, going so far at times as to finance operations on an equal footing with the European Union (EU). The market integration of agriculture – the perfect illustration of unequal development ranging from intensification to the abandonment of land use – disqualifies the farming profession from claiming, in the name of a homogeneous occupation of the land, that it is the social mainspring of the rural world. Even where this control of the land remains decisive – which is the case of the areas targeted by the agri-environmental programmes – the other users of nature not only want the legitimacy of their relationship to be recognized but also focus their criticisms of agricultural practices on the very basis of their justification, namely their guiding technical and instrumental rationality. Such a confrontation involves more than a mere spatial distribution of the different social uses, or more than the conception of a mode of production more friendly to the natural environment. Through these questions, or beyond these questions, the issue raised is that of rurality and the agrarian heritage as the archetype of a certain way of living together.

*And Modern*

The fact that negotiation is one of the principal vectors for collective construction, the fact that it is also based on heterogeneous perceptions of nature, and consequently subject to a series of formalizations, places the agri-environmental question in a debate of central importance to modern societies. The uncertainty prevailing over the convergence between agricultural practices and the constructed data of nature has shown how negotiation could be an instrument facilitating public management provided it is representative of local diversity and that the ensuing compromise is synonymous with commitment and adhesion. Inventing a 'negotiated' form of development was the goal of these operations which set out to found productive practices on social compromises based on locally produced heterogeneous knowledge.

The entire question is to know how the mechanisms adopted will fit in with the forms of joint management and, more generally, with local life: do they draw up a new action-oriented body creating obligations between social groups and, in this respect, organizing social life in its relationship with a given territory? Do they carry a design for the rural world which, less focused on agriculture alone, will convey a more faithful image of its own sociological diversity? The agri-environmental programmes have generated action groups whose outcome was the invention of new social references with a view to the organization of space based on environmental objectives. As such, they forced the rural actors to analyze their social relations on the basis of their relationship with nature. Whether such a relationship is legitimate and forms part of the common good, by a justification of a patrimonial or ecological nature, is not indifferent for the resulting social configuration; but, in both cases, it marks a break with a view of the rural world which, through its privileged reference to farming, presented a model of a relationship with nature and development based on a rational management of technology. This model is faced with a crisis: the arena of the negotiation is, among others, a place where a less agricultural rurality – and, in this respect, a new type of rurality – is being developed.

**Technical Democracy and Local Democracy: a Tense Relationship**

The implementation of 'negotiated development' deserves, however, a more discriminating analysis particularly when viewed from three angles of which we shall illustrate the one we consider to be the most important. Firstly, owing to the very complexity of the problems treated and the rather diversified configuration of the actors involved, implementation expresses itself by having recourse to expert opinion and knowledge in order to provide a basis for public action. This link between knowledge and action presupposes that the decision-making process will be clarified by the mobilization of science or by having recourse to expert opinion. Thus, the state of knowledge, with its batch of uncertainties and disagreements, limits the field within which the political choices or compromises are made. But commitment and adhesion, to which the decision-making process must lead, cannot be satisfied with a primarily scientific approval, especially when this approval has been the subject of contradictory debates and fails to enjoy an image of absolute rigour. Where we could have expected a strict framework for behaviour created by the scientific background provided by the experts, the uncertainty that hampers action (which is all the greater in the event of controversy between scientists) gives a major role to negotiation between the actors in the confrontation of the different points of view. To ensure these points of view lead to individual commitment and collective action, the expert opinion must not contradict the expectations of any of the parties but contribute to the construction, which is necessarily temporary in the event of disputes, of the agreement.

The second problem raised in a 'negotiated development' process is the fragmentation of the public space resulting from the creation of these forums where negotiation is organized around the resolution of technical and territorial issues. These forums are fragile to the extent they create, for the duration of the negotiations, impermanent territorialities or 'collectives' destined to disappear once the problems have been resolved. In the agri-environmental operations, there is a sort of instrumentalization of these forums in favour of a management policy for the territories of which the State and the EU retain responsibility. In other words, these forums produce rules of an essentially temporary nature and without immediate incidence on the sharing of responsibilities. It is not by chance that supervision and possible sanctions, in the event of a failure to respect the clauses of the agri-environmental contract, are never subject to debate within the forums themselves. But, at the same time, they offer a form of organization and regulation of life in society which appeals to a certain notion of modernity based on procedural democracy: the rules of the game are first derived from extremely localized arrangements. Is it the very nature of these partial and highly detailed arrangements that explains that the social agents involved in these forums are unable to escape from the collective framework developed by a technical solidarity or an 'ecological interdependence'?[9] Our hypothesis is that it is precisely on the relationship between a technical democracy, proper to forums such as these, and a local democracy, established by virtue of citizenship, that depends a possible reconstitution of rural societies. It is a question of seeing how this instrumental approach to the solution of technical or environmental questions contains a global collective project and, as such, is liable to offer a basis for agreement between social groups different from the traditional modes of governance. The aim is to study the ability of rural societies to integrate socio-technical arrangements such as these, founded on expertise and negotiations, in their own mechanisms of local democracy.

The third problem raised by 'negotiated development' includes the first two because it refers to the representativeness to be attributed to such forums in their capacity as 'parliaments'. Who speaks? Who represents whom? Who is excluded, on the side of nature or from the point of view of society? Because, in the event of a deficit in representation, is not the negotiation perverted and transformed into a 'masquerade' (H.-P. Jeudy, 1996)?

And yet, these mechanisms clearly participate in the recomposition of the local rural arena. To illustrate this proposition, we will refer to a case study, that of a negotiation system for the drafting of the terms of an agri-environmental programme in marshlands on the Atlantic coast. This system gathered together most of the actors of the rural space, from institutional representatives of the farming profession to local politicians and nature protection associations. When we compare the profile of the negotiation after a period of five years, we notice that the focus of the negotiation has shifted considerably, and this coincides with a different composition of the 'parliament' (Table 8.1). What happened exactly? And why?

Initially, the discussions focused on the place of local stockbreeders in terms of their access to contracts. Farmers living on the 'high land' neighbouring the marsh in fact, chiefly use the water meadows of the marsh. In the negotiation, of which the goal rapidly became that of determining the role of ecological criteria on which to base an agri-environmental policy, the local politicians assumed the role of spokespeople for the marsh stockbreeders, thereby seeking to impose equivalence between local origin and service rendered to the environment. An activity disqualified owing to its being controlled by farmers from outside the marsh and to its lack of economic viability, wetland stockbreeding saw itself granted a patrimonial and domestic legitimacy by virtue of the relationship to the environment it is supposed to maintain. In the first 'parliament', there was little question of ecology properly speaking but of the right to be represented or not – as expressed by the administrative term of 'eligibility' – in the forum of agri-environmental negotiation (Billaud, 1996).

Five years later, the second 'parliament' presented a completely different aspect. The negotiations had shifted their focus, either because of the questions being debated, or from the point of view of the representativeness of the forum. These changes cannot be attributed to the effects of the contractual policy implemented until then:[10] in the case studied, 38 per cent of the farmers who signed a contract lived in the marsh in 1992, but only 32 per cent in 1997. In other words, the renewal reveals a marginal deterioration in the presence of marsh-based farmers in the operation. And yet, a comparative examination of how the agri-environmental system was put together clearly shows that the question of the presence of the stockbreeders from the marsh is no longer a subject of debate, but the major problem is that of water and, through it, the ability of a discussion forum to find a compromise between social development and the reproduction of the wetland ecosystem. One of the significant changes involved in this shift in focus is the disappearance of the local politicians from the second parliament and the arrival of the stockbreeders themselves, at times representing more than one half of the participants at the consultation meetings. To this, we should add the arrival of new 'natural entities' of which the representatives of nature protection associations make themselves the spokespeople. But the latter do not enjoy a monopoly of this discourse since new actors also join the arena (the Water Agencies, for example) whose presence is required for the satisfactory inclusion of the materiality of nature in the negotiation process. In other words, around the question of representativeness, i.e. the composition of the 'parliament', the development options can be singularly different.

This example allows us to illustrate more precisely the tension that characterizes the relationship between the technical democracy created in these forums and the traditional local democracy of rural societies. The chart allowing a comparison between the two 'parliaments' of the agri-environmental operation (see below) illustrates what, from the development point of view, the exercise of a technical democracy can imply since, in our opinion, the mechanisms involved in local democracy still had a considerable influence on the 1992 negotiations owing to the preponderant part played by rural politicians. We shall focus on four aspects enabling us to characterize technical democracy on the basis of this case study.

*Justification Integrating the Point of View of Expert Opinion*

It is clear that the position of those who held the 'natural' point of view is no longer, in 1997, the one observed five years earlier: the legitimacy of their participation in a negotiation, fiercely contested by the farmers' representatives, is now accepted. This change is related to the proximity created by all negotiation processes: people learn to get to know one another. But, above all, it is related to the position that the associations can now assume by making ecological criteria a veritable subject of social debate; in doing so, they acquire legitimacy as bearers of expert knowledge. This set of criteria is much less subject to the uncertainty that affected it earlier: there is no more waiting for biological samples which cannot keep pace with the administrative deadlines for compiling a file; the representatives for the 'natural' point of view are better equipped to answer the questions put by the other partners when required to give a precise description of an ecology-based technical reference system. The technical translation of the ecologist's point of view is based, for example, on the contribution of precise knowledge on nesting periods, by the proposal of minor water-control structures making it possible to retain the water on certain meadows, etc.

**Table 8.1  Comparison between the two agri-environmental 'parliaments'**

| | 1991 – 92 | 1997 – 98 |
|---|---|---|
| Initial phase | Overall territorial diagnosis (followed by) Professional register of grievances | Emergence of a question: aquatic environments and water management (followed by) Identification of problems not included within the frame-work of agri-environmental negotiation |
| Expression of the problem | Environment = Maintaining the marsh stock-breeder | Environment = Maintaining the natural meadow, maintaining the stockbreeder, modes of water management |
| Legitimate problems | What stockbreeders? Conflicting options: - no selection - membership (local/profession-nal) - ecological zoning (followed by) biological statement | What topics? - pasture (brambles and thorn bushes) - thistles/mosquetos /pond turtles - natural character of the meadow - bacteriological pollution - quality of the 'drops' - reeds and sparrows - open pools (jas) and network |
| Technical debates regarding the eligibility of the parcels | What practices on the parcels? - date of mowing - stocking rate per hectare - winter grazing | What practices for maintaining the ditches? - initial state - date, frequency and type of clearing - treatment of clearing pro-ducts |
| Priority in the attribution of the financial envelope | Marsh and subsequently 'non-marsh' areas | Natural meadow (with adjustment depending on the percentage of Natural Mea-dow/Surface area) and subse-quently temporary meadow ' |
| Semantics of the contract | 'strong/weak commitment' | simple/reinforced biological contract' |
| Experts | ADASEA, Chamber of Commerce, INRA, LPO | INRA, LPO, UNIMA, Water Agency, EID, Plant Protection Department of the DDA |
| Dominant actors | Local politicians (mayors) | Stockbreeders and experts |

*Source:* Author's inquiry.

*A Socio-technical Statement*

In a certain manner, this second phase in the agri-environmental negotiations offers a clear illustration of a certain recalcitrance on the part of nature: it finally imposed itself in the construction of the negotiation mechanisms through the mobilization of questions related to water levels, the maintenance of ditches and water-control systems. The new terms to sanction this inclusion of the marsh ecosystem provides a new syntax for the human aspect – the marsh stockbreeder rather than the native farmer – and seeks to draw up rules guaranteeing the relationships between the stockbreeders and their natural environment. It goes without saying that an approach of this kind makes the basis of the agreement much more complex: the equivalence between the stockbreeder and the environmental question is founded on a system of inter-relations which mobilize agronomic practices, the natural meadow as an agro-system but also the management of water. It stands therefore as a socio-technical statement to the extent it tries to harmonize the relationship between nature, technology and society.

*A Forum Comprised of Human and Non-human Actors*

All this was only possible at the price of a significant modification in the actors' respective roles or, at least, in their distribution. The stockbreeders imposed themselves in the technical discussions (to the detriment of the local politicians, as mentioned above) but new partners also emerged. By calling on their know-how, the negotiation forum tried to integrate problems for which the stockbreeders acted as spokesmen, such as the elimination of mosquitoes (by a semi-public body), the treatment of thistles (by a specialized department of the Departmental Board of Agriculture). The same demand to contain the human and the non-human within the scope of a concrete assumption of the problems, such as the management of water between different administrative units (whence the inclusion in the discussion of the 'Departmental Plans for the Development of Water Management') or pollution between catchment basins (whence the presence of the Water Agencies). In other words, the extension of the contributions of the actors with, as its corollary, the marginalization of certain partners of the first phase (the local councillors) leads to a significant extension of the issues, the integration of new questions and, consequently, of new topics[11] where the boundary between what concerns the social dimension and what concerns the natural dimension is increasingly difficult to perceive.

*A Compromise Structured in Relation to the Topics (both Technical and Natural) in Question*

The fact of focusing on the question of water and, more precisely, on the issue of the maintenance of the ditches, offers a clear illustration of an approach tending to integrate, by the creation of new and complex relationships, issues that concern the functioning of the ecosystem and issues that are dependent on the functioning of the stockbreeding system: reeds and sparrows; *jas* and connections to the ditch, etc.

The fact that the highlighting of this inter-relation shifts the heart of the negotiation, previously focused on the parcel of marshland, to the ditch surrounding it, is a sign of the increase in complexity. The ditch is a place both for the circulation of water but also for social practices related to its maintenance. Unlike the parcel of meadow which is only a matter for a private and univocal practice (stockbreeding), the ditch evokes several practices (stockbreeding but also hunting and fishing) as well as the interdependence proper to the circulation of water in a hydraulic system where private and collective concerns are mixed. This is why it forms the conclusion of an attempt to reach a translation of the different points of view in both technical and social terms. But the maintenance of the ditch is normally a matter for the traditional contract between the stockbreeder, the owner of the ditch and the marsh association. Against the administration (which is opposed to its inclusion), the partners in the negotiation, from the stockbreeders to the nature protection associations, make the maintenance of the ditch a condition of the agri-environmental contract. This reveals, not only an approach founded on vested interest but the determination to make the contract the basis of a new pact between the marsh stockbreeder and the wetland. In a sense, the accent placed on the ditch and its maintenance translates a quest for equivalence, on the part of the actors in the consultations, between the 'needs of nature' and the 'needs of individuals'.

Thus, these systems – which we have described as 'forums' (in their ability to reconcile data of a hybrid nature) and as 'parliaments' (owing to the importance of their representative nature from the point of view of technical democracy) – represent the stakes of a collective experiment to translate technical points of view into social development. But the fact that an experiment such as this supposed that the local politicians, the archetypical representatives of local democracy, play a secondary role to the stockbreeders and experts, who have in fact forged a new alliance within the public discussion, is not unimportant. There undoubtedly exists a veiled conflict of representation and know-how, divergences regarding the place of different actors in development and in the justification of the common good. On the one hand (local democracy), the boundary between nature and society is drawn explicitly; there exists a clear determination to keep them separate and so to protect the human and social dimension (in this case, the native stockbreeders) from the constraints of nature (by virtue of which stockbreeders not living in the marsh may be legitimate); on the other hand (technical democracy), there is an attempt to do away with, or at least to blur, the boundary between nature and society, to develop a compromise based on a hybridation of representation which, in fact, disqualifies what is at the very basis of rural legitimacy, namely membership of a local community. It is obvious that a development process, liable to take charge of environmental issues and, as such, to be in line with sustainability will be caught between these two extremes which put pressure on contemporary rural societies and play a decisive role in social change.

**Agriculture and Rurality Faced with 'Negotiated Development': a Few Thoughts by Way of Conclusion**

'Negotiated development' corresponds, when it lays the foundation for public policy, to the irruption of a certain type of 'modernity' in rural societies.

For agriculture, it represents a new technological challenge as conveyed by the notion of 'high-tech farmer', advocated by the OECD. In a model of this type, information – i.e. the cognitive integration of the new goals of sustainable development into practical know-how – becomes an input as vital as agrochemicals. After productive modernization, are we moving towards an ecological modernization?

From this point of view, it cannot be denied that farmers are now faced with an additional uncertainty concerning the socially acceptable goals of agriculture: the legitimacy related to the production of food, which was at the centre of post-war agricultural modernization and which justified a demand for parity with other social groups, is not longer sufficient. Society expected from the farming industry what was required from the other major industrial sectors of the nation: to feed the country and then the world: the link between production and work was immediate. The widening of the notion of agricultural and food production to the idea of a manager of space and service-based agriculture began several decades ago through the socio-economic changes in the profession (emergence of multiple activities, for example) and in society as a whole (development of 'green tourism', etc.); what is more, a large number of farmers have long anticipated these changes, either by economic opportunity or through a distancing with a 'productivist' model. The agri-environmental policy confirms these processes and their diverse approaches and, in a certain manner, creates space for discussion about the validity of agricultural co-management, with those who were shunned because they denounced, for example, the environmental damage of a certain model, finding a certain audience in these new arenas: the press has long, and increasingly, provided an echo for the internal tensions running through the majority of farmers' professional bodies.

As soon as the legitimacy of agricultural work is no longer founded on the mere accumulation of goods produced in a space essentially considered as a support, as soon as the subsidy mechanism introduces a disconnection, an 'uncoupling' between revenue (direct aids per farm have risen from FRF22.000 to FRF110.000 in the space of 5 years) and production, the result can only be a blurring of the professional terms of reference in agriculture.

What the agri-environmental experience suggests is that the construction of professional identity will no longer be the exclusive preserve of organizations involved in a dialogue with the State. This identity is also being forged in these 'local operations' during which the farmers are involved to extremely varying degrees and appropriate (or not) the standards proposed for another attitude towards nature.

The agri-environmental measures, by making local arenas emerge from negotiations about agricultural goals and practices, by proposing a system of individual contracts, marks to a certain extent the appearance, or the return, of the farmer as a subject-actor. In this sense, the construction of a professional identity in agriculture will undoubtedly follow more complex routes or, in any case, paths that are more difficult for observers to decipher, to the extent this construction will occur at the limits of social, but also territorial, membership.

This notion of membership, which is one of the key elements in social and professional identity, lies at the heart of a large number of debates to the extent we are currently observing a revision of the traditional modes of governance and a rise in 'socio-technical systems' founded on the mobilization of expertise in public action. At the scale of rural societies, the adjustment of these systems with the traditional mechanisms of social representation is a burning question for all democratic processes and, more precisely, for the rurality of the future.

These forums, which call increasingly on new systems of obligation between the social actors – cf. the former agri-environmental contracts in Europe and the 'territorial exploitation contracts' in France today – represent different ways of organizing life in society, which illustrate an idea of modernity founded on a procedure-based democracy, i.e. rules of the game which are not derived from a general principle but from local arrangements.

These new systems of action and co-ordination visible in the socio-technical systems linked to the environment are creating a local arena which more accurately reflects the sociological diversity of the rural world, where the farmers not only are no longer alone, but must accept other legitimacies for managing space. The problem of these systems is that they are fragile (linked to public policy procedures) based on a territorial dimension that may prove to be ephemeral (based on the identification of an environmental problem: the case of a catchment basin, for example): their logic is first of all instrumental and is not presented as the test of a collective project expressing shared values. Acting together is not enough to make a collective action. What rural sociability is conveyed by such negotiation procedures? The question remains unanswered.

Two recent events in France show how much the stakes of this new form of development may lead to a particularly uncertain recomposition of local rural arenas. With the application of the 'Habitats Directive' (Natura 2000) there emerged a coalition between the traditional actors of rural space (farmers, hunters, fishermen, foresters, landowners, etc.) built on the opposition to the scientific expertise of the naturalists; the large number of contradictions within this 'group of 9', namely 9 organizations representing the rural actors, were erased during this revolt which led, in 1996, to the suspension of the application of the procedure by the government at that time. More recently, these same 'rural actors' turned against their pig breeders who were demonstrating vociferously against the collapse in prices: they then resorted to being consumers, acutely aware of the quality of products but also of their living environment and, in this respect, they broke the image of a faultless rural solidarity around their farmers. This reaction is quite unusual and shows that agriculture no longer has the same legitimate foundation within rural societies. The environmental problems and, more generally, the issues highlighted by sustainable development can generate (owing to their wide scope ranging from nature to styles of life) contradictory reactions among the rural actors.

The reconstructions triggered in rural territories by the inclusion of these questions must be monitored in the future, particularly from the point of view of the new forms of democratic procedures to which they may lead. The whole question is to know if this 'negotiated development', an emblem of the irruption of modernity in rural societies, announces the demise or the rebirth of these societies.

## Notes

1  France was recently sanctioned by the Council of State because of failure to respect the 4-month deadline for the local negotiation of the Habitats Directive (*Le Monde*, October 6, 1999).

2  This expression (*'démocratie technique'*) is borrowed from an article by Callon (1998).

3  We shall use several terms to describe these bodies created around a consultation system set up to implement a public development policy. The 'local arena' (*'scène locale'*) refers to a public debating area, as defined by H. Arendt (1986) and, after her, by J. Habermas (1987). The notion of 'forum', that we shall use later in this text, refers to more technical arenas such as agri-environmental steering committees; it is related to the networks of knowledge and action for which M. Callon and P. Lascoumes (1997) have proposed the expression 'hybrid forum'. And, lastly, we shall also use the term 'parliament' when deciphering the organization and semantics proper to forums of this kind; this rather evocative term was used by B. Latour (1991) in his survey of the Moderns, with respect to what he calls the 'parliament of things'.

4  Switzerland, for example, where incentive policies related to taxation and labelling are the most advanced, although the negative impact of agriculture on the environment is generally less dramatic than elsewhere: it is well known that, owing to a constitutional system where each citizen is entitled to have recourse to a referendum, consumer pressure can be stronger in Switzerland.

5  'It seems clear that sustainable agriculture is not a return to modes of production with low technological input. The farmer must become a "high-tech" manager capable of unravelling the complex inter-relations between technology and the environment to obtain acceptable yields and profits while reducing the ecological impact of his activity and pursuing predetermined objectives in terms of the environment and public health' (OECD 1994: p. 13).

6  In other words, 'an optimization of the efficiency of the use of inputs so as to avoid unacceptable levels of pollutants' (OECD 1992). Another example of what is currently at stake in ecological modernization is provided by the USA where ten federal agencies are working on pilot programmes devoted to sustainable agriculture.

7  Thus, the representative of the United States admitted before the other OECD countries that 'opinions diverge on these questions, depending on the people who express them. For example, in the case of pesticide residues and the contamination of underground water, there exists proof supporting the existence of a serious problem but there are just as many who deny this argument' (OECD 1994, p. 220). The Franco-British conflict about the exportation of British beef provides another, extremely current, example of a controversy about risks.

8  The term is borrowed from N. Dodier (1997), but his work focuses on the industrial environment.

9  The expression is borrowed from M. Mormont (1996) who uses it to describe 'the

constraints of the biophysical world in the social universe'.
10  It should be remembered that the local agri-environmental operations had to be renewed every 5 years.
11  This is the case of the *box tortoises*, for example, which were included in the negotiation in favour of including the ditches in the negotiation: to determine the date for cleaning out the drainage ditches, they played the same role as the growth of grass in relation to mowing times, in the first-phase contracts; this is also the case for the *jas*, immersed areas situated on the plots of land and inherited from the morphology of the salt marshes which also refer back to the question of the level of water in the ditches with which they interact, etc.


## References

Arendt, H. (1986), *Vies Politiques*, Gallimard, Paris.
Billaud, J.-P. (1995), 'Agricultura Sustentavel nos Paises Desenvolvidos: Conceito Aceito e Incerto', *Agricultura Sustentavel*, no. 2, pp. 23-32.
Billaud, J.-P. (1996), 'Négociations Autour d'une Nature Muette. Dispositifs Environnementaux dans les Marais de l'Ouest', *Etudes Rurales*, no. 141-142, pp. 63-83.
Billaud, J.-P. (1997), 'Les Opérations Agri-Environnementales: Quelques Réflexions sur la Gestion des Espaces Agricoles et Naturels', *Comptes Rendus de L'Académie D'Agriculture Française*, no. 83, pp. 27-35.
Boltanski, L. and Thévenot, L. (1991), *De la Justification. Les Economies de la Grandeur*, Gallimard, Paris.
Callon, M. (1998), 'Des Différentes Formes de Démocratie Technique', *Annales des Mines*, pp. 63-73.
Callon, M. and Lascoumes, P. (1997), 'Information, Consultation, Expérimentation: les Activités et les Formes D'Organization au sein des Forums Hybrides', Proceedings of the seminar of the Programme *Risques collectifs et Situations de Crise*, Ecole Normale Supérieure des Mines, Paris.
Dodier, N. (1997), 'Remarques sur la Conscience du Collectif dans les Réseaux Sociotechniques', *Sociologie du Travail*, no. 2, pp. 131-148.
Habermas, J. (1987), *Théorie de L'Agir Communicationnel*, Fayard, Paris, 2 Vols.
Jeudy, H.-P. (1996), 'Le Tout-Négociable', *Autrement*, no. 163, pp. 20-30.
Latour, B. (1991), *Nous N'Avons Jamais été Modernes. Essai D'Anthropologie Symétrique*, La Découverte, Paris.
Latour, B. (1999), *Politiques de la Nature. Comment Faire entrer les Sciences en Démocratie*, La Découverte, Paris.
Millon-Delsol, C. (1993), *Le Principe de Subsidiarité*, PUF, Paris.
Mormont, M. (1996), 'L'Environnement entre Localité et Globalité', in M. Hirschhorn and J.M. Berthellot (eds), *Mobilités et Ancrages*, L'Harmattan, Paris.
OCDE (1992), *Séminaire sur les Technologies et Pratiques d'une Agriculture Durable*, OECD Publications, Paris.
OCDE (1994), *Pour une Production Agricole Durable: des Technologies plus Propres*, OECD Publications, Paris.

Chapter 9

# Market Liberalization and Rural Development Policies: Two Faces for the Future of the Italian Countryside

Francesco Di Iacovo

## Introduction

The aim of this paper is to outline how Italian agriculture and the Italian countryside are adapting to changing social demands, a question that Agenda 2000 is also concerned with. A negotiating process between both old and new objectives and subjects which are involved in the political arena of European agriculture is required to carry out this programme successfully. It is well known that Agenda 2000 represents a compromise between European and international views on policy. It is a starting point for defining new opportunities for the European agricultural model. Manifold problems, underlined not only by Agenda 2000 but also by numerous innovative actions, have led to the creation of this document. The document tries to give response to four different questions.

The first question regards intervention in the agricultural markets. The current liberalization of exchanges concerns two issues: the reduction in degrees of market protection and the establishment of proper hygiene regulations with the application of voluntary certification standards. Both attempt to promote a controlled deregulation of the markets, especially in consumer relationships. It may be possible to identify the above two issues as the hard core tendency of European Union (EU) policies, which have been imposed, in part, by commitments from within international circles.

The second question regards the European agricultural response to the current diversification and specialization trends in consumer models in rich countries. Increased attention to products of superior quality leads to more lucrative consumption. This question is a very important one for Mediterranean countries that have all along kept agriculture and the traditions of the countryside alive before modernization techniques and processes began wiping out diversity.

The third question concerns environmental qualifications of productive processes. This matter takes on different aspects depending on whether it is applied to models addressing intensive production or to models dealing with abandoned rural areas.

The fourth question regards the strengthening of intersectorial relations among the agricultural, rural and urban worlds; policies that (in EU terminology) are often defined for the development of rural areas but which (at least as far as Italy is concerned) could, just as well, apply to certain urban poles of industrial development.

The first two questions concern two different types of farming practice and different ways of interpreting the relationship between the subjects in the chain and the consumers. In both cases the market represents the ultimate instrument used to deal with the final consumers. On the other hand, co-ordination between those involved in the productive phases can take place through the market or, alternatively, by a resoldering of community links within local systems.

The last two questions address the connection between agriculture and the collective needs. From this point of view, the rural world's capacity to supply environmental services (all the while preserving roots and traditions that make a local community different) and to ensure a high standard in the quality of life depends on the management of the political network.

Agenda 2000 defines two different pillars for the Common Agricultural Policy (CAP): the policies for the markets and policies for rural development. Both are expected to have a strong impact on the evolution of Italian agriculture.

In order to present the Italian case and the paths of the impending reorganization, this paper has been organized as follows. First a reflection on the European agricultural policies that have been adopted until now is presented; then a summary of some of the productive and organizational elements that characterize Italian rural areas; and concludes with a picture of what is happening now followed by an analysis of the possible impact that the new political and socio-economic agenda will have on Italy.

## The CAP Evolution

The CAP has always been known to be faithful to the principal elements of European agricultural growth. Like the changes undergone by industrial production, the start of the CAP has promoted a shift from an artisan-like system in agriculture to a more industrialized one, thereby facilitating the rise of Fordism and mass production.

Food shortage was the springboard for these policies. The main objective was to modernize the Agro-Alimentary System (AAS) and integrate such industrialization as had already occurred.

By the same token, at the start of the 1970s and during the 1980s, when industrial development policies were showing the first signs of crisis, the over-production of agricultural goods was then again regulated by intervention. This took the form of a kind of redundancy fund extending throughout the agricultural sector and activated as a result of the redistribution of resources from the expanding industrial and tertiary sectors.

This phase comes up against with the onset of significant intrasectorial and territorial disparities in European rural areas. The geographical CAP impact, the

EU financial crisis, as well as international instability led to a rethinking of agricultural policy. The crisis of a model of relations that was used to regulate agriculture in the 1960s and 1970s was analogous to the goings-on of the industrial sector. Around that time in the non-agricultural sectors, a disparity began developing between workers who consolidated their own rights and the growing number of people excluded from the job market and often deprived of some of their rights to receive assistance. In agriculture, while some farms enjoyed an increasing ability under the CAP regulations to gain access to EU funding, at the same time a growing number of small farms were not managing to obtain sufficient subsidies, especially those in mountainous and marginal areas.

The start of a third phase, that of liberalization and the conversion of the primary sector and territorial readjustment, takes place under the MacSharry reform. Agenda 2000 represents the evolution and the synthesis of this new compromise. It aims to favour a regulated transition toward a system that was created to protect interests that have become increasingly difficult to defend, with policies capable of offering the European Union's citizens new benefits. This amounts to a third route between established protectionism and outright liberalization. It also constitutes the basis for the creation of new regulations meant to avoid distortion in a market which is progressively opening up, and increase distribution of services by the rural world as it involves itself in the development of European society.

Policies regarding the development of rural areas, therefore, become ideal instruments for the relaunching of the European society model and a means for improving the quality of life for EU citizens. This applies not only for active policies regarding the environment; the EU, starting with rural development policies, introduces themes and ways of working that are relevant to the subject of development in a more general sense, along with the co-ordinating instruments that help to promote it (Tarditi, 1997).

What remains to be assessed is to what extent public intervention will provide a new kind of buffering zone and whether it will be capable of generating relationships and new opportunities for areas that have until now been marginalized from applied models.

As far as the changes that are taking place are concerned, it would be wrong to read Agenda 2000 without reflecting on those policies able to influence (possibly through generating further confusion) the productive relations of European rural areas. Particularly important in this respect are the impact of co-operation and closer ties between the world's macro regions (Mercosur, South Mediterranean, Peco) on competition in the markets for primary foodstuffs; the effects of modifications in the horizontal regulation of foodstuffs and the diffusion of voluntary certification on quality in agro-alimentary production; and in a strictly social field, the repercussions of active policies against unemployment and their reflection in the agricultural labour market (especially in the Mediterranean), in relation to the difficult problem of migratory flows.

The agricultural model of reference at the start of this new century distinguishes itself from those that have for many years dominated the CAP horizons. It reflects an evolution in the demand for foodstuffs and a growth in environmental awareness (a growth in sensitivity towards the environment). More generally, to a greater extent, however, these developments pave the way toward a new organization of exchanges on a global scale and an entirely new attitude toward synergies among the primary sector, rural areas and development.

Agenda 2000's reference points can be encapsulated as follows: a more self-regulated market, a better response to consumer needs, and a greater coherency as far as the requirements of local systems of development are concerned. These three, seemingly conflicting, points reflect the new organization of productive relations on a global scale.

The globalization of the economy, carried out as a result of a continual opening of exchanges, is causing a shift of wealth on a world scale. Those benefiting from this redistribution are not only nation-states but, above all, those local systems that are capable of actively participating in the control of current economic and political dynamics (Amin and Thrift, 1994; Budd, 1998; CNEL, 1996; Di Iacovo, 1994). In such a transformed situation, the European Union can continue to compete on the market for primary products by means of a consolidation of processes that have become efficient through introduction of advanced technology; at the same time on a local level adopting systems with high levels of structural flexibility and a capacity to handle change dynamically.

Faced with shrinking protection from the market, local systems can rise to the new challenge of increasing the competitive advantages of the farms that work within them. The local system, with its social, productive, institutional, and environmental structure, becomes a new organizational paradigm for active participation in a globalized society (Dansero, 1996; Dei Ottani, 1995).

Within local systems, agriculture is obtaining new impetus for various reasons. First, in agriculture (and as is often the case in the industrial sectors) the affirmation of a dense network of relations between members facilitates management of innovation, increases the productive flexibility of the system and favours the creation of an atmosphere in which information can circulate. Secondly, agriculture is becoming rejuvenated. The efficiency of a local system increasingly depends on the capacity to handle the delicate balance between natural resources, human settlements and productive activities. This agricultural activity can be decisive in distributing various types of services that are useful in promoting efficiency of the system itself. All along, agriculture can continue to contribute to the development and efficiency of local systems, to raising levels of intersectorial and social cohesion and to favouring consolidation in terms of quality of resources available all the while attending to the environment (De Benedictis, 1998; Russu and Di Iacovo, 1997). Finally, the progressive opening up of the markets broadens the consumer target. This element, especially in the case of fully developed productive systems, permits the opportunity for operating in rich consumer markets thus satisfying a new and widening range of needs (Polidori, 1997; Sylvander, 1996).

European farmers find themselves up against a very different configuration of relationships – in the management of productive processes but also in the relationships between rural and urban areas – in spite of their very mixed origins and various backgrounds (Struffi, 1993). It is due to this very diversity that each national reality, and especially every local system, starting out with its own resources, follows paths of development that go hand in hand with its local needs. Consequently, it is not a coincidence that the EU favours the principles of subsidiarity and co-financing. The EU seems intent on ensuring a basis of common references in terms of macro-objectives and instruments, leaving it up to the individual national and local decision-making bodies to adjust their policies in accordance to local developments.

## Features of Italian Agriculture

Italian rural areas have not steered completely clear of difficulties during the evolution of European Community policies. The first phase of the CAP in Italy coincided with an economic boom which would have enabled Italian production to compete with the world's leading economies. From the 1950s to the 1970s a basis was formed for the transition and commodization of Italian agriculture that would end up determining the onset of numerous territorial disparities and types of organization (for some features of Italian agriculture see Tables 9.1-9.2).

**Table 9.1 The main structural features of Italian agriculture**

| Year | People employed ( ,000) n° | Farms (,000) n° | Farm land (,000 hectares) | Arable land (,000 hectares) |
|---|---|---|---|---|
| 1950 | 8.610 | 4.500 | 27.000 | 18.900 |
| 1960 | 6.118 | 4.294 | 26.572 | 18.600 |
| 1970 | 3.605 | 3.607 | 25.065 | 17.491 |
| 1980 | 2.760 | 3.269 | 23.631 | 15.843 |
| 1990 | 2.070 | 3.023 | 22.702 | 15.046 |
| 1996 | 1.403 | 2.800 | 22.500 | 15.000 |

*Source*:  STAT. The numbers of farms and hectares in 1950 are estimated; in 2001 the census scores 2.611.000 farms.

**Table 9.2 The main economic features of Italian agriculture**

| Year | (prices in current billion lire) | | | (prices in constant 1996 billion lire) | | | |
|---|---|---|---|---|---|---|---|
| | Gross production | Inputs | Food Consumption | Gross production | Inputs | Food consumption | Self sufficiency |
| 1951 | 2.565 | 313 | 3.431 | 59.950 | 7.316 | 80.191 | 74.8% |
| 1960 | n/a | n/a | 6.153 | n/a | n/a | 111.820 | 0.0% |
| 1970 | 6.741 | 1.501 | 14.463 | 83.716 | 18.641 | 179.610 | 46.6% |
| 1980 | 30.983 | 8.807 | 57.434 | 103.530 | 29.429 | 191.915 | 53.9% |
| 1990 | 58.138 | 16.798 | 156.371 | 77.306 | 22.336 | 207.927 | 37.2% |
| 1996 | 69.933 | 19.735 | 204.000 | 69.933 | 19.735 | 204.000 | 34.3% |

n/a:  not available

*Source*:  ISTAT.

This period was characterized by: strong migratory movements within the country from southern to northern areas and from mountainous areas to the plains, leaving behind the marginalization of entire territories (Tellarini, 1992); by an internal reorganizing of the existing entrepreneurial framework and an important transference of assets from this sector and, by the continuing presence of a large number of small enterprises, very often run on a pluriactive basis.[1]

The high number of enterprises show how in Italy, perhaps more than elsewhere in the EU, land management is often a completely different matter from participation in the real economic dynamics of the primary activity. The Italian agricultural sector can be divided into two types: enterprises that, although organized in different ways, are integrated into international, national and local markets (Cannata, 1989; Cannata, 1995; Fabiani, 1991; Nomisma, 1994; Nomisma, 1996; Favia, 1992; Montresor, 1999); and organizations whose link with the productive world has waned over time, but which, although often deprived of a real economic capacity owing to their essentially subsistence character, have become significant from a social standpoint as well as in terms of earth management.

At the same time, the CAP has encouraged specialization of production on a territorial basis allowing for a differentiation between:

- areas that are strongly integrated into national and international markets;
- plains areas with a pronounced tendency toward industrial, zootechnics and grain production (mainly concentrated in the North of Italy and around Rome);
- areas of irrigated plains with a strong proclivity toward flower, fruit and vegetable activities (mainly concentrated in Emilia Romagna and in reclaimed areas of the South);
- areas which are highly dependent on EU intervention; hilly areas and dry plains with a tendency toward grain production (situated in central and Southern Italy);
- areas engaged in specialized consumer markets; hilly areas with a strong inclination towards arboriculture (vines, olive trees, orchards) (predominantly situated in central and Southern Italy, but in the case of vine hills and orchards, in the North too);
- abandoned areas; mountainous and upland areas of limited productive potential.

Areas tightly integrated into the markets and those that do not depend in any way on EU policies are often to be found in economically developed environments or in areas in which the agricultural economy has a strong foothold. The areas of arboriculture exist in economically dynamic settings. In Italy any consideration of the future of rural areas entails dealing with a wide spectrum of productive contexts, interdependent sectors and, last but not least, social actors. For this reason the future of Italian rural areas is inseparably linked to both sides of the equation; namely, liberalization of markets and policies of rural development.

*A Few Reflections on Organizational Models of Italian Agriculture*

The need to tackle the modernization of agriculture and the specific conditions of Italian production – from a viewpoint at once territorial, social, and institutional – has favoured the growth of different organizational models, and above all a powerful structuring of relations in accordance with a strict sequential logic, the growth of agro-alimentary districts, and the definition of production models paying due consideration to local values.

The exclusive reference to the paradigms of modernization has often been seen in terms of close supervision of the agricultural productive units and a progressive integration, initially through the markets and ultimately by means of contracts along the various phases of the chain. Though linking them to the market, this kind of integration released the enterprises from their territorial components.

This kind of organization was logically predicated on a continuous amplification of the productive scale large enough to make it possible to introduce technological innovation in the form of fixed capital. While conducive to enhancing the enterprise's and the agricultural entrepreneur's image (they were no longer farmers but entrepreneurs), this model ended up isolating the enterprise from the territorial context, linking it, in terms of representation, to the main national trade unions and to centres of political power, especially to the majority party.

What is involved here is a model badly suited to the Italian agricultural world, with its characteristic obstacles of inflexibility in the land market and high population density. Thus a new innovative organizational pattern began to develop in Italy in parallel with the modernization process, its objective being to weave a dense web of relations between the corporations and between the public and private sectors. It was a flexible form of agricultural specialization and, particularly in certain areas it has through its distribution of services promoted dissemination of leading technologies within the corporation itself (which would ordinarily not have access to them because of the smaller size of the business). This is an organizational model for agriculture well known in Italy and it is referred to as the 'rural district' model (Becattini, 1987; Cecchi, 1992; Fanfani and Montresor, 1998; Iacoponi, Brunori and Rovai, 1995; Montresor, 1999).

On the basis of such principles, numerous agri-alimentary districts were established, usually specializing in just a few products (Parmesan cheese, Vingnola's cherries, Fucino's horticulture, meats from Modena, etc.) which have been able to consolidate their positions on the market because of the creation of strong synergies on a local scale (Nomisma, 1996).

Italian law has recently recognized the importance of the 'rural district' model. Its role is to improve the efficiency of the local systems and to promote the co-operative attitude among different actors at the local level. This last-mentioned factor is a very important way to promote multifunctionality in agriculture.

Neither of the schemes described entails explicit ties with territory. Only in areas where modernization of agriculture was carried out with some difficulty was there a strengthening of the link between agricultural production and the character of the locality. This choice, which was initially vestigial and rendered obligatory by a number of different considerations, then proved to be compatible with the evolution of consumer models from mass production toward differentiated commodities.

The major components of this third model are: the creation of a productive climate sensitive to the values of the local culture; the revitalization of traditional artisan productive techniques of the area; the strengthening of relations between all the economic and social actors operating the microchains, and even employing associative tools (the strong symbolic link between the product and the attractions of the area (landscape, traditions, culture) – a good innovative both in the management of the image of the product being sold on the markets and in the creation of an ambient market space.

The products are, thereby, commercialized by means of local markets or a more direct management of relations between production and consumption. This productive model has strengthened links between product value and value of locality, employed the attractions of regions and increased its touristic appeal. What is evident is that here are strategies that find their impetus in the decisions of the EU regarding the introduction of Protected Denomination of Origin (PDO) and Protected Geographical Indication (PGI), and even in obstacles that follow the bureaucratic organization of the procedures.

The logic in the co-ordination of this organizational system is partly similar to those of the district, though they offer increased access to intersectorial dynamics. The success of these systems lies in the capacity to adapt modern technologies to the problems that are encountered, and to manage the innovations linked to the certification and marketing of the products. These undertakings, taking up positions along short-lived chains and niche markets, also make alliances with the world of traditional distribution possible, supplying those structures that by this point have been made obsolete by organized distribution with the chance to seek a better position.

## The Impact of Reform on Italian Agriculture

On account of structural, organizational, environmental, and territorial diversities of Italian rural areas, the effects of Agenda 2000 differ according to case. In this case too we will try to highlight the impact of the reform, taking into consideration the theme of the liberalization of the markets, and the unfolding of integrated rural development with the aim of diversifying productive systems, revitalizing life conditions in marginalized areas where development is to take place, favouring the production of environmental services and increasing relations of functionality between areas of urban and rural development.

### *The Joys and Pains of Market Liberalization for the Italian System*

In the near future a progressive liberalization of the primary foodstuffs markets is expected. This expectation is taken into consideration in every approach to political reform. A progressive liberalization of the markets could lengthen the main chains, giving a more important role to transformation phases and OD. Besides the effects of competition, liberalization will also have effects on the relations installed along the chains. As far as Italian agriculture is concerned, there are two possible routes forecast; the search for increased competitiveness by way of reduced costs, and the search for more fluid organizational relations along the chain or within the agroalimentary districts.

The first aspect often comes up against objective problems regarding the production context (environmental, technical, etc.). The second is very closely related to the enterprises' ability to carry out the co-ordination phases thoroughly amongst themselves and with subjects beyond the chain, whether or not they are present in the territory where the productive organizations operate.

It would be wrong to think that in a more open market competitive advantages derive only from the price of the primary materials made available. From the buyer's viewpoint this assessment can apply only for the total cost of the supplies. This cost is made up of two parts: the price of the primary materials purchased and the costs of the transaction itself. This implies that even in the primary materials market fresh products (and basic converted products) are competitive factors stemming from certain parameters, namely, product reliability, respect for hygiene and health standards, and the management of logistics. One of the most urgent problems derives from the need to make the individual participants in the chain jointly responsible, thus making it easier to obtain products.

The EU's employment of a set of regulations strengthening hygiene and disease prevention introduces the concept of correct hygienic procedures, providing for the use of voluntary certification instruments, and tending to encourage transparency and self-control.

Competition is managed paradoxically by moving away as much as possible from purely market-based relations, adopting those co-ordinating instruments that can lower the costs of the transaction, and reduce the costs of control together with those costs that derive from the purchase of products that subsequently turn out not to conform to regulations. In this respect the productive world is required to establish more direct and transparent relations along the chain.

This approach requires considerable effort from agricultural establishments, especially in what concerns those products whose supply markets continue to be their community partners. This is particularly true in Italy,[3] where the transformation industry and organized distribution have by now consolidated relationships with the import markets.

The reference points for the reorganization of agricultural markets are: reduction in production costs, reduction in transaction costs and improvement in liaison along the chain. All these choices not only affect the commodities markets but could also be extended to circuits that make a stand for clearer features of quality.

Differentiation in consumption makes for continuous distinctions between commodities and quality markets. Nevertheless, for some product types the boundaries are insignificant and fluid.

The consumer's final perception of quality is often conditioned by the degree of refinement of the communication instruments used. It is therefore easy for a product subject to voluntary certification (whether of production processes or of the product itself) to enter into competition with a product certified PDO or PGI. Even for those circuits that develop around superior quality products (meats, oil, fruit and vegetables, wine, etc.) it would therefore be unthinkable to disregard the organizational evolution taking place in the markets for key products.

*Paths of Integrated Rural Development*

As far as rural development policies are concerned, the EU points out numerous functions that rural areas can be called on to implement in order to requalify development paths (Andreoli, 1996; Andreoli, 1998; Marsden, 1993). It is well known that the EU tends to distinguish between various types of rurality in accordance with the kind of link or the distance in relation to the poles of urban development. This is a particularly significant distinction for Italy where the problems of the rural world can vary significantly even between areas that are not a great geographic distance from each other.

Thus, in Italy, under the heading of integrated rural development, we can include methods (Basile and Cecchi, 1997; Brunori, 1994; Cecchi, 1998; Di Iacovo 1996; Osti, 1996; Russu and Di Iacovo, 1997; Saraceno, 1993):

- which aim at revitalizing and diversifying the economies of regions which have until now been excluded from economic development (mountainous and hilly areas of the hinterland);
- which aim at requalifying the presence of agricultural activity and its function as a distributor of environmental services within local systems;
- in areas where the latter has lost the required impetus for an industrial and tertiary system;
- in areas where such intensive processes have consolidated to such an extent that excessive pressure is generated on natural resources.

The common denominator in each of the aforementioned work hypotheses is the reference to relations over a territory and to a qualification of the life standards – albeit economic and/or environmental and social – of local productive systems. A change which must not be taken too lightly; which implies the definition of a way to achieve regulation at a local level appropriate for the framework of resources and limits – local or not – to which reference is made.

The indirect effects of the application of fourty years of intervention policies in agricultural markets were, in Italy, above all in weaker areas, the loss of entrepreneurial ability in a generation of farmers and the indifference of the political classes and local councils toward problems in agriculture and rural areas.

The CAP extended the role of the bureaucracy at national and EU level, and the role of markets and clan relations were likewise organized into national pressure groups for lobbying. On the other hand, the CAP helped to streamline relations in local communities. In this way the pronounced decisional centralization, often limited to confrontation between EU institutions and member states and national representatives of professional organizations led to the exclusion of local representatives from decision-making. The regulation of support assured by means of a market forged by strong EU intervention helped to free producers from the constraints of relations within the territory, but, at the same time, ended up greatly limiting the dialogue taking place among themselves and with local government. This system eroded the rural community's chances of drawing up projects consistent with the requirements of the territory.

Nowadays, therefore, the inauguration of locally determined development plans implies an intense revision of the interpretation of the categories of 'reciprocity', 'organization', 'market' and 'political exchange'. Negotiation at a local level has once again become important, as has the understanding between entrepreneurs and the system of local representation. Public administration and the structure of local institutions represent an important key to the understanding of the difficulties and potential success of development carried out locally.

This is true both when we speak of the environmental impact of intensive production and when we deal with requalification in the running of territory in areas subject to exodus and abandonment, and in addition to this when programmes for diversification or expansion in local economic systems are introduced. In Italy there are already concrete examples in all three indicated directions that will be summarized to provide material for discussion

The first has its origins in a model from Emilia Romagna, employed to detect and reduce the environmental impact of fruit and vegetable production. The parallel application of regulations and incentives for the introduction of techniques which would have a weaker environmental impact ended up by generating process innovations (in the introduction of techniques of greater compatibility) and product innovations (creating a brand of products with greater guarantees). The assumption behind this was the confrontation between the world of productive associationism (endowed with high technical professionality and considerable ability to operate even in OD markets) and the regional political administration (in its turn efficient in the management of relations with administrations outside the regional boundaries – state and EU – and in the dialogue with subjects working within the regional territory).

The co-ordinating instruments used are based on a mixture of organization, local bureaucracy, bargaining forms and the management on the market of new products. This local instrument emerges as an organizer and distributor of new services.

The second is represented by the attempt launched in Tuscany[4] in planning of the structure and management of rural areas. Here, as in many other areas of Italy, economic development has generated a rift between rural and urban areas and a marginalization of high-altitude areas. The concentration of civil and industrial growth along the valleys of the Arno River has aggravated the subalternity of the neighbouring rural areas, which have often become recipients of urban and civil waste and inert supply points.

Choices have generated conflict in areas that are witnessing a renewed emphasis on the values of rurality. The break-up of an equilibrium between mountainous and valley areas has caused water problems and persistent damage to material resources and human life, and has provoked doubts about the very viability of some productive districts.

In this region the management of soil plays a fundamental role in assuring the vitality of local systems. At the same time the development of rural areas provides the key to the creation of employment in activities of great cultural, social and environmental potential.[5]

With intent to assuring a better-balanced management of rural spaces, a revision of local autonomy regulations has been carried out using criteria of decisional subsidiarity involving various institutional levels (local councils, communities of local councils, provinces, and regions) in the definition of the use of territory and in the analysis of its resources.

A plan is still being carried out, and has met with considerable difficulties, for reorganization of a web of environmental services in the territory beginning with already existing agricultural activities. The same Regulation, 2078/92, has been widely implemented but has had very few practical effects. The promotion of services for environmental management has proved problematic. Inside the public administration, the staff is not used to combining the themes of territorial planning with those of development. Agricultural trusts also have many internal difficulties in their relationships with the public staff and their members. The new environmental and rural policy approaches necessitate new attitudes; the new political course calls for a quick passage from a demanding attitude to a more proposal and project-oriented one, both in the relationship with the members and with the public staff. From this point of view the first steps in the definition and the application of the Rural Development Plans (Regulation 1257/99) do not seem to be introducing a deep change. In the end, the farmers too have to work actively in order to rebuild a dense network of relationships among themselves and with the other institutional and private actors at a local level. The last is a necessary strategy for the reconstruction of a common vision – even technical – between a multitude of owners who are interested to a different extent, and often hardly at all, in the management of the land.

The third example relates to methods for strengthening and diversifying the economy that are underway in many rural areas, thanks to EU instruments such as (IC LEADER and Structural Funds) but not only these (Di Iacovo, 1996). They derive inspiration from the National Parks' system. The actions concentrate on strengthening local values, attributing increasing value to traditional organization of rural settlements, availing themselves of artistic centres, of the system of old farms and specific configurations of the landscape.

This kind of organization makes use of rural spaces through thematic activities (horse-riding tracks, wine, oil) and is supplemented by productive activities and rural hospitality. As far as the Parks are concerned, the activism of the public has favoured, albeit not without conflict, the revitalization of entire areas and the involvement of local populations in development programmes with the potential to make use of the resources that the Park intends to save.

Whether or not this kind of initiative can get off the ground is often contingent on the outcome of a local negotiation aimed at reaching agreement on a development opportunity. This is often a lengthy task, whose difficulty lies in dealing with the forms of co-operation and conflict that crop up with local subjects and their delegations. As far as the Parks are concerned there is a clear trade-off between the values of conservation and those of creating development opportunities. In other cases, such as in the LEADER initiative, the grant of public contributions is assured in exchange for the commitment to adopt work methodologies in which a public-private decision is the central node for the construction of a directive for the development of a territory.

Partnership is an instrument that can be applied in Italy even outside rural areas in matters related to industrial development (territorial pacts, area contracts, etc.) (CNEL, 1996; De Rita, 1998; Dente, 1992). Where there are no subjects acquiring legitimacy through their adoption of specific regulations (for example outside LEADER areas), the task of managing the territory can absorb considerable resources and time, particularly in a situation where the world of public and private delegates is richer, and there is time pressure to start up co-operation between the sectors. In any case it is clear that we must redefine from the start the mechanisms of accumulation and distribution of economic and political advantage that gain strength from the local system.

A more concrete problem that is difficult to solve arises where there is a pronounced decline in human resources, especially in areas that are remote from centres of development and where nothing is being done that will provide hope for sudden changes in direction. Low levels of efficiency and poor management of money are also likely in areas where social cohesion is slow to appear.

## Conclusion

The models of reference for agriculture at the beginning of this new century present very different features from the one that has long dominated the CAP horizons. Globalization of the economy, brought about through continuous opening up of exchanges, produces reallocation of wealth on a world scale. The risks of this process are linked to the exclusion of a large part of the producers from the market, including at the local level.

Not only have some nation-states benefited from this redistribution, but local systems have also shown themselves capable of actively participating in the control of the economic and political dynamics. The reference points of Agenda 2000 can be summarized in three slogans which are, for a number of reasons, sometimes in conflict; more self-regulated markets, more attention to consumer needs, and more congruence with the need for the development of local systems.

In such a transformed situation European agriculture can continue to compete on the commodity material markets in two ways; through the consolidation of processes that have been made efficient by the introduction of high technology, or, through the adoption on a local scale of systems of great structural flexibility with a sound capacity for dynamic management of change.

Faced with a diminished protection of markets, local systems take on a new role, that of increasing the competitive advantages of the enterprises that operate in the local system. The local system, with its social, productive, institutional, environmental structure, becomes a new organizational paradigm for an active participation in the dynamics of a society that is opening up more and more.

Within the local systems agriculture acquires new roles, for various reasons. The first is that, as with what happens in industrial districts, the establishment, in agriculture, of a dense network of relations between subjects facilitates the management of innovation, increases the productive flexibility of the system, and favours the creation of an atmosphere in which valuable information can rapidly circulate.

The efficiency of a local system can be measured in accordance with its capacity to manage the delicate balance between natural resources, human settlements and productive activities. Agricultural activity can be decisive in the distribution of various types of services which could increase system efficiency. At the same time, however, agriculture could contribute to the development of the local systems, increasing levels of intersectorial and social cohesion favouring their consolidation in terms of the quality of resources, including environmental, that are available.

Moreover, the progressive opening-up of markets widens the range of different consumer types that can be reached. This is a circumstance that, especially in the case of developed productive systems, creates opportunities to focus on rich consumer targets, satisfying a wide new array of needs.

In the Italian case more could be done to empower rural areas, such as introducing new organizational settlements on a local scale. This should be the first step towards active participation in the global as well as the local market, producing environmental services on a local scale and promoting integrated rural development approaches.

The creation all over again of an organic social block, often starting with very different participants and interests (Marsden et al., 1993) is a prerequisite for embarking on the consolidation of local economic systems (and potentially entire rural areas). Nevertheless, the risk is that, faced with the real difficulty of making an economy out of rural areas, structural intervention ends up becoming a way of stabilizing the loose dynamics of very weak territories, while simultaneously justifying the lack of commitment on a national level. What remains, in any case, is the chance to fully exploit a favourable moment to reaffirm the role of rural areas in Italian society.

## Notes

1   Ninety six per cent of the Italian farms present in the EU observation field are family farms. They represent 34 per cent of EU farms. Only 12 per cent of them are greater than 12 EDU in size. Approximately 800.000 Italian farms receive the CAP contributions.
2   In the South of Italy Commercial Macro Organizations are being set up (in the sectors of oil, market gardening and pasta cereals) for the purpose of creating and consolidating relations in local and regional chains.
3   A region where residential settlement is under way throughout its territory, enjoying a good income and benefiting from a growing appreciation of the positive aspects of rural

life on the part of an increasing number of tourists and/or inhabitants of more recent arrival.

4 Over the past few years, youth employment in agriculture has increased as a result of the introduction of financial incentives and of reconsideration of the quality of life available in rural areas.

## References

Aglietta, M. (1979), *A Theory of Capitalist Regulation: The US Experience*, NLB, London.

Amin, A. and Thrift, N. (eds) (1994), *Globalization, Institutions and Regional Development in Europe*, Oxford University Press, Oxford.

Andreoli, M. (1996), 'Rural Development and Structural Funds: Between Market Forces and Public Intervention', in M.E. Furlani de Civit (ed.), *Development Issues in Marginal Regions II: Policies and Strategies*, Ex-Libris, Mendoza.

Andreoli, M. et al. (1998), 'From Economic Marginality to the Problems of 'Quality of Life', in *Dynamics of Marginal and Critical Regions*, Proceedings of the International Geographical Union, Coimbra.

Andreoli, M. et al. (1996), 'Processes of Change in Marginal Area Family Farms: the Case of Garfagnana', in M.E. Furlani de Civit (ed.), *Development Issues in Marginal Regions II: Policies and Strategies*, Ex-Libris, Mendoza.

Basile, E. and Cecchi, C (1997), 'Differenziazione e Integrazione nell' economia Rurale', *Rivista di Economia Agraria*, no. 1-2.

Becattini, G. (ed.) (1987), *Mercato e Forze Locali: il Distretto Industriale*, Il Mulino, Bologna.

Brunori, G. (1994), 'Spazio Rurale e Processi Rurali: Alcune Considerazioni Teoriche', in Panattoni, A. (ed.), *La Sfida della Moderna Ruralità*, STAR, Pisa.

Budd, L. (1998), 'Territorial Competition and Globalization: Scylla and Charidis of European Cities', *Urban Studies*, Vol. 33(4).

Cannata, G. (ed.) (1989), *I sistemi Agricoli Territoriali Italiani*, Franco Angeli, Milano.

Cannata, G. (ed.) (1995), *I sistemi Territoriali Agricoli negli Anni '90*, Arti Grafiche La Regione, Cosenza.

Cecchi, C. (1992), 'Per una Definizione di Distretto Agricolo e Distretto Agroindustriale', *La Questione Agraria* no. 46.

Cecchi, C. (1998), 'La Ruralitö nella Periferia e nel Sistema Locale', *Aestimum*, no. 2.

CNEL (1996), *Laboratori Territoriali. Competizione e Leadership nella Questione Settentrionale*, Roma.

Dansero, E. (1996), *Eco-sistemi Locali*, Franco Angeli, Milano.

De Benedictis, M. (1998), 'La Qualitö Agro-ambientale: Problemi e Politiche', *La Questione Agraria*, no. 70.

De Rita, G. and Bonomi, A. (1998), *Manifesto per lo Sviluppo Locale. Dall' azione di Comunitö ai Patti Territoriali*, Bollati Boringhieri, Torino.

Dei Ottati, G. (1995), *Tra Mercato e Comunità: Aspetti Concettuali e Ricerche Empiriche sul Distretto Industriale*, Franco Angeli, Milano.

Dente, B. and Gario, G. (eds) (1992), *Il Governo Locale Possibile*, Franco Angeli, Milano.

Di Iacovo, F. (1994), 'Le Politiche Agricole e lo Sviluppo Locale: Note a Margine di una Ricerca sulla Diffusione del Set-aside', in A. Panattoni (ed.), *La Sfida della Moderna Ruralità*, STAR, Pisa.

Di Iacovo, F. (1996), 'Institutions and Guidelines of Adaptation in Rural Areas: the Example of Tuscany', in M.E. Furlani de Civit (ed.), *Development Issues in Marginal*

*Regions II: Policies and Strategies*, Ex-Libris, Mendoza.

Di Iacovo, F. et al. (1996), *L' Esperienza Leader in Toscana: la Rivitalizzazione delle Aree Rurali per la Crescita dell'Economia Regionale*, ARSIA, Regione Toscana, Firenze.

Fabiani, G. (ed.) (1991), *Letture Territoriali dello Sviluppo Agricolo*, Franco Angeli, Milano.

Fanfani, R. and Montresor, E. (1998), 'Istituzioni ed Imprese nel Percorso di Sviluppo dei Sistemi Locali di Produzione Agroalimentare', *La Questione Agraria*, no. 69.

Favia, F. (1992), 'L' Agricoltura nei Sistemi Produttivi Territoriali', *La Questione Agraria*, no. 45.

Favia, F. (1995), 'Sui Distretti Agroalimentari: dal Prodotto al Territorio', *La Questione Agraria*, no. 57.

Garofoli, G. (1992), *Economia del Territorio*, ETAS Libri, Milano.

Iacoponi, L. et al. (1995), 'Endogenous Development and Agricultural District', in J.D.Van Der Ploeg and G. van Dijk (eds), *Beyond Modernization. The Impact of Endogenous Rural Development*, Van Gorcum, Assen.

Marsden, T. et al. (1993), *Constructing the Countryside*, University College London Press, London.

Montresor, E. (1999), 'I Sistemi Locali di Produzione Agricolo-Alimentare', in CNEL (ed.), *Rapporto 1998 sull'agricoltura. Schema di Osservazioni e Proposte sul Tema: L'agricoltura tra Locale e Globale. Distretti e Filiere*, Annali CNEL, Rome.

Nomisma (1994), *Rapporto 1994 Sull'Agricoltura Italiana*, Bologna.

Nomisma (1996), *Rapporto 1996 Sull'Agricoltura Italiana*, Bologna.

Osti, G. (1996), 'Nuove Filosofie di Sviluppo delle Aree Rurali Italiane', *Aggiornamenti Sociali*, XLVII, no. 9-10.

Osti, G. (1997), 'Il Contadino Postmoderno. Valori e Atteggiamenti degli Agricoltori Italiani', *Sociologia Urbana e Rurale*, no. 54.

Perulli, P. (ed.) (1998), *Neoregionalismo*, Bollati Boringhieri, Torino.

Piore, M.J. and Sabel, C.F. (1984), *The Second Industrial Divide: Possibilities for Prosperity*, Basic Books, New York.

Russu, R. and Di Iacovo, F. (1997), *Il Territorio e lo Sviluppo Rurale nella Politica Comunitaria Europea*, Accademia dei Georgofili, I Georgofili, Quaderni, VIII, Firenze, pp. 133-212.

Saraceno, E. (1993), 'Dall' Analisi Territoriale Dell'Agricoltura allo Sviluppo Rurale', *La Questione Agraria*, no. 52.

Struffi, L. (1993), 'Ci sono Ancora Differenze tra Campagna e Città, in Italia, per quanto Riguarda i Valori?', *Annali di Sociologia-Soziologisches Jahrbuch*, Vol. 9(I).

Sylvander, B. (1996), 'Normalization et Concurrence Internationale: la Politique de Qualite Alimentaire en Europe', *Economie Rurale*, no. 231.

Tarditi, S. (1997), 'L'Italia di fronte agli Orientamenti della Nuova Politica Agroalimentare Comune', in Proceedings of the XXXIV SIDEA Meeting, Torino.

Tellarini, V. ( 1992), 'Some Questions about Socio-Economic Marginality in Rural Areas of Developed Countries', in O. Gade (ed.), *Spatial Dynamics of Highland and High Latitude Environments*, Appalachian State University, Boone, North Carolina.

Chapter 10

# Transformations of the CAP and the Need for Reorganizing Agricultural Policy in Greece[1]

Michael Demoussis

## Introduction

The pursuit of the aims of the Common Agricultural Policy (CAP) (agrarian development, adequate living conditions for farmers, balanced markets, the pursuit of self sufficiency in agricultural products and reasonable prices for the consumer), agreed upon at the 1957 Treaty of Rome, is based on the well known principles for the operation of the CAP (unified markets, community preference, economic solidarity), which were themselves set in 1958 and made more specific in the 1960s with particular institutional measures for the production and trade of agricultural products.

The operation of the CAP is based primarily on two categories of measures, the first of which is composed of immediate price support measures for agricultural products, which, depending on the product, include price guarantees and intervention, export subsidies, import duties and a production aid package. These measures absorb approximately 90 per cent of the European Union (EU) resources for the agricultural sector. The second category, that of structural measures, deals mainly with the factors of production (measures for regional development, farm and crop restructuring, extensification of production, early retirement plans, income supplements, environmental actions, agricultural product processing and marketing, etc.) and absorb the remaining 10 per cent of the agricultural budget.

The implementation of the CAP over the past decades has had very specific results. The productivity of agricultural inputs has increased substantially, producers' incomes have been maintained at satisfactory levels and sufficient food supplies have been secured for the EU as a whole. At the same time, significant imbalances have been created in most markets of agricultural products (surplus production), agricultural inputs have not been used in the most effective and efficient way, the prices at which European consumers can purchase foodstuffs and agricultural products are much higher than international prices and, finally, the viable and integrated development of the countryside is delayed. Furthermore, the direct expenditures of the European Agricultural Guidance and Guarantee Fund

(EAGGF) increased from five billion ECU in 1975 to forty-one billion ECU in 1997, while the allocation of these expenditures has been uneven, since 20 per cent of EU producers receive 80 per cent of the provided total assistance. In addition, the indirect and invisible cost of the CAP is estimated to be close to fifty billion ECU.

Surplus production has turned the EU into a particularly aggressive exporting power through the use of export subsidies, a development that has prompted reactions from other exporting countries and especially from Third World countries which depend on the exports of agricultural products for their development. In early 1980s, the results of the CAP application were questioned by EU countries, especially by those with a small agricultural sector but a disproportionately heavy burden from the CAP implementation; and by consumers who were forced to pay high prices for agricultural products and food supplies.

These reactions in turn caused many adjustments and re-evaluations of the CAP. Indicative of these are reforms such as, the co-responsibility levy, the introduction of stabilizers (production and quota thresholds), early retirement schemes and direct income supplements. In addition, during the General Agreement on Tariffs and Trade (GATT) negotiations (Uruguay Round), the EU proceeded to a re-evaluation of certain Common Market Organizations (CMOs) and was forced to make significant compromises, with a period of implementation from 1995 to 2000, such as: 20 per cent decrease in the overall support of the agricultural sector, 36 per cent decrease in import duties, 36 per cent decrease in export subsidies and 21 per cent decrease in the volume of subsidized exports. The EU has withstood constant pressure from the international community to remain fully oriented to the gradual reduction, and even to the abolition, of subsidies, in particular, and of support for European agriculture, in general (Maraveyas, 1992b; Tracy, 1997).

## The Effects of the CAP on Greek Agriculture

Before Greece's accession to the EU in 1981 and in anticipation of the accession, the agricultural policy implemented by the Greek government was very similar to the CAP of the time. With the admission of Greece into the Common Market, the Agricultural Fund undertook the relevant costs of essentially the same agricultural policy and the national resources that were freed, as a result, were channelled into other sectors (Demoussis et al., 1988; Georgakopoulos et al., 1985). Today, the CAP contributes to the creation of almost 50 per cent of Greece's total agricultural income, making the agricultural sector the most heavily supported sector of the Greek economy; the national and Community funds devoted to Greek agriculture stand at 67 per cent of the agricultural sector's gross added value in market prices (1995). Of course, this incredible support is not distributed uniformly to the various regions of the country. For example, support provided to the farmers of Thessaly is almost double the national average, while support given to the farmers of the Aegean islands is less than half of that average. As a result, the farmers of Greece have ceased to constitute the poorest income class and their combined income from all sources (agriculture, pluriactivity of the head and other members of the household, rent and interest income, transfer payments, etc.) matches, and in many cases supersedes, that of non-farming households.

Given the persistent structural problems of the Greek agricultural sector (small and fragmented land holdings, an elderly farming population with low level of education, the bad state of the co-operative movement, the bureaucratic public administration, etc.) and the decreased efficiency of the implemented CAP, it seems only natural that the competitiveness of Greek agriculture has not improved. Contributing to this development is also the fact that Greece, since its admission to the EU, has consistently exhibited its own inability to institutionally intervene with effective national reforms in areas that were not directly in the narrow interests of the CAP (such as land use, environmental protection, vocational training and education of farmers, co-operative movement, etc.).

Compared to other EU countries, Greece has the smallest size of farm holdings, the smallest size of farmland per farmer and the second lowest added value produced per farmer. In addition, for every size class of holdings, Greek farmers use about the same amount of labour but definitely less capital and intermediate inputs compared to farm holdings in other EU countries. Despite this, Greek farmers utilize available resources in the most efficient way enjoying higher incomes than the farmers of comparable size holdings elsewhere in the EU.

The relevant statistical data for the implementation of the CAP in Greece show that: a) total agricultural income exhibits a cyclical diachronic trajectory, which has, until now, ensured stability (increasing in the1980s and decreasing in 1990s), b) agricultural income per fully employed labour unit has increased dramatically since the 1980s and has tended to stabilize in 1990s, c) investments in the primary sector have remained at low levels, mainly because of their conventional nature, which seems to have exhausted its potential, and d) the demographic problem remains serious, despite the recent concurrent improvement, which is characterized by an increase in the absolute number of young farmers and a clear trend of farm restructuring exhibited by young farmers in favour of large holdings and against small.

## Major Aspects of Greek Agriculture

*The Structure of Landholdings*

The average size of Greek farm holdings is the smallest in the EU (between 3.5 and 4.3 hectares) and has not increased notably over the past thirty years. However, when the average size of farm holdings is measured in economic units, (one economic unit was 1.200 ECU in 1993), then it shows a substantial improvement of almost 100 per cent increase in the period from 1983 to 1993. The small size is typical also of livestock-breeding holdings, with the difference that the average number of animals per holding increased significantly between 1980 and 1993. Furthermore, according to the latest available data, almost one-fourth of farm holdings cultivate land of less than one hectare and only 8.5 per cent land over 10 hectares.

Between 1961 and 1991, the proportional participation of very small and of large farm holdings increased (Damianos et al., 1998). In spite of this increase, the participation of small, in terms of economic size farms, has decreased since 1980, while the participation of holdings of over sixteen economic units has tripled. Young farmers (those under thirty-five years of age) cultivate larger holdings (in both hectares and economic units) than those cultivated by members of other age groups. The fragmentation of landholdings remains a serious problem, despite voluntary efforts and mandatory redistribution. The average number of parcels per holding in 1993 was around 6.5 and almost 20 per cent of holdings are composed of ten dispersed parcels. Irrigated land takes up today almost 30 per cent of total cultivated land, a proportion which places Greece before all other EU countries, and particularly those of the Mediterranean basin like Portugal (23 per cent), Italy (19 per cent), and Spain (11 per cent).

*Cereal and Vegetable Production*

Cereals, the most land-consuming crop, taking up about 40 per cent of all cultivated land, exhibit a declining trend as a result of price policies followed after Greece's accession to the EU. The exception to this trend is the cultivation of corn that has been expanded because of the widespread use of hybrids and the increase in irrigated lands, which have resulted in a substantial increase in yields. With regard to industrial crops, cotton tripled its cultivated area between 1971 and 1995. Today it occupies almost 40 per cent of the country's irrigated land as a result of its incorporation into the corresponding CMO. Based on the proposals submitted until now, the future of cotton production seems ominous, given the anticipated further decreases in support. Tobacco, after a steady fall connected with price policies and restructuring programmes, shows signs of stabilization in both production and size of cultivated land. However, further decreases are anticipated for low-demand varieties (eastern varieties) and for certain low quality varieties (Virginia). Furthermore, a decrease in the number of tobacco farmers and a stabilization of incomes at today's mediocre levels is anticipated. The cultivation of sugar beets has stabilized since the early 1990s and the present production level corresponds to the capacity of the existing processing plants. Sunflower cultivation has decreased primarily as a result of the extended drought and the significant difficulties encountered in the marketing of the product.

The contribution of orchards and vineyards is steadily decreasing within the grain/vegetable production of the country as a result of major restructuring programmes. Most of the decline can be attributed mainly to the decrease of Corinthian grapes. Such quantitative trends, however, should not hide significant qualitative shifts that have already started to appear with the increased planting of high quality orchards (replanting with a different type), the production of high quality wine and of organic grape products. The reform of the wine CMO is expected to have positive implications for the product and its quality level.

Areas cultivated with trees have increased significantly. The most visible increase was observed in olive tree cultivation because of a substantial increase in the support level for olive oil. Nuts and seeds have been promoted with a relatively successful restructuring, although Greece has yet to realize its comparative advantages for the off-season production of quality fruits, nor has it managed to organize conditions of transport and marketing of these products to an acceptable level. The restructuring of the CMO for fruits and vegetables seems likely to increase the pressure on the producers of low-quality products. Finally, the organization of the off-season vegetable production has not been completed as yet, mainly because of organizational and institutional weaknesses of marketing and promotion, which have not helped to take advantage of the potential of the domestic and the EU markets (Damianos et al., 1998).

## *Meat and Dairy Production*

Meat production of all kinds has increased since the accession of Greece to the EU. More visible has been the increase in white meats (pork, poultry) because of the possibilities of industrialized production and the disconnection of production from cultivated land, which has been a limiting factor in the past. Despite this, it is estimated that poultry production will not continue to increase, because of structural weaknesses in the poultry sector. The production of meat has increased mainly because of the very effective support programmes, but in recent years, a significant downgrading of the health status of the animal stocks has been observed. The increase in meat production, however, was much smaller than the comparable increase in demand, a situation that resulted in increased meat imports, which have more than tripled in recent years. Today, the value of meat imports represents almost half the value of domestic production.

Dairy production (cow, sheep and goat) has also increased and the problem of imports is not as serious as in the case of meat production, despite the appearance in the Greek market of new products such as highly pasteurized milk. Further increases in milk production are not easy to come by given the structural problems of Greek agriculture, the problem of shrinking pasture land, in terms of both quantity and quality, and the absence of the necessary technical infrastructure for the organization of sheep and goat herds (Damianos et al., 1998).

## *Fisheries*

Greek fisheries make up approximately 0.25 per cent of the Greek Gross National Product (GNP) and about 2 per cent of the agricultural GNP. The sector provides employment to 30.000 fishermen in inshore, open-sea and overseas fisheries. From 1983 to 1993, the fishing fleet showed significant increases in the number of vessels and installed horsepower (HP), while the Gross Registered Tonnage (GRT) capacity of the fleet showed a small decrease. The increase in the number of vessels is attributed mainly to the inshore fleet. It is worth mentioning that despite efforts to regulate and control production, total fisheries output grew by 70 per cent in the period 1983 to 1993. This increase is attributed primarily to the inshore fisheries and to a lesser extent to open-sea fisheries. Finally, fish consumption increased from 140.000 MT in 1983 to 240.000 MT in 1993. Seafood imports increased from 40.000 to 60.000 MT over the same period, while an impressive increase in the production of farmed fish was also observed. At the end of 1996, 226 aquaculture units were in operation, with a total production level of approximately 22.000 MT composed mainly of sea bass and sea bream.

*Forestry*

Forests do not constitute an important sector of the Greek economy in terms of their contribution to GNP and employment, since the gross product of this sector does not exceed 1.5 per cent of the agricultural GNP. Forests however, take up 19 per cent of the land area of Greece and together with 'forested areas' (partial tree coverings and grazing lands) they constitute the single most important land use. Two-thirds of forests belong to the state and produce almost 2.5 million cubic metres of wood, covering 30 per cent of the country's needs. The most pressing problem of the sector is the ownership regime of forests and 'forested areas', which is characterized by permanent disputes between the state and third parties. If the ownership problem is dealt with efficiently and promptly then with the completion of the forest registry even the serious threat of forest fires could be minimized. With proper management, forests can contribute to the preservation of the environment and the development of mountainous areas providing the necessary additional job opportunities and incomes to those still living in there.

*Food Industry: Food Supplies – Alcohol – Tobacco*

The food-alcohol-tobacco sector is among the most dynamic industrial sectors of Greece. It depends directly on Greek agriculture, which is, for the most part, the sole supplier of the basic inputs. It accounts for 21 per cent of the total employment of the Greek industry, 28 per cent of gross production, 26 per cent of added value and almost 30 per cent of gross investment. The significance of the sector becomes even greater when one takes into account the multiplier effects of it in the Greek economy as a whole. The food sector has the third highest income and employment multipliers and the eighth highest product multiplier of the thirty-five sectors in the Greek economy, while alcohol and tobacco show a comparable performance. Most industrial establishments in this sector process agricultural products locally, processing that can be either very elementary e.g. selection and packaging, or very complex e.g. production of prepared food. It is worth mentioning that at present there are in operation 815 milk processing units, more than 100 fruit and vegetable canning factories, 3.270 olive presses, 300 small wine factories and a plethora of small industrial food processing establishments for almost every Greek agricultural product.

*The Regional Dimension of Greek Agriculture*

Greek agriculture prior to its accession to the Common Market was characterized by a regional specialization in the composition of agricultural output, determined primarily by the different natural, economic and social conditions prevailing in the different parts of the country. After the accession, various regions of the country began to specialize in the production of the products with the highest support, increasing in this way their dependence on the CAP. The ongoing developments in agricultural policy are expected to have national and 'region specific' effects, which will have to be kept in mind when planning the orientation of Greek agriculture. A second regional dimension of Greek agriculture concerns the mountainous and less favoured areas widespread in every region of the country. Mountainous and less favoured areas have specific agricultural characteristics and a particular socio-economic environment, which determine to, a large extent, their development potential. The regional differentiation of Greek agriculture constitutes one of its most basic features that must be given special emphasis in the course of planning the Third Community Support Framework (Third CSF).

**Further Restructuring of the CAP**

In view of: a) the new round of World Trade Organization (WTO) negotiations, b) the restructuring of agriculture in Eastern Europe, c) the enlargement of the EU, and d) the increased emphasis on environmental issues and problems, it is clear that the EU will have to proceed with further transformations of the current CAP (Sarris et al., 1994; Ostrom, 1995; Maraveyas, 1992a). The 'messages' from the ongoing political discussions indicate that these adaptations and transformations will proceed along four axes:

a. A rapid convergence of EU to international agricultural product prices will be pursued and international prices will constitute the main long-run determinant of EU agricultural production.
b. For those EU farmers who will be judged as needing continued protection of their income under the new regime, support will take the form of direct income supplements, under the condition that this support will be accompanied by a change i.e. a reduction in the use of agricultural inputs.
c. Efforts will be made to ensure that further increases in productivity, resulting from technological progress and scientific research, will not be accompanied by an increase in the volume of production. At the same time, a decrease in the use of inputs such as agricultural land, fertilizers and pesticides will be encouraged.
d. Agricultural policy will soon place greater emphasis on environmental goals and EU farmers will be compensated when they take actions that promote the production of public goods that protect and conserve nature and the environment.

Agenda 2000, with an implementation period of 2000 to 2006, constitutes, in essence, the preparatory stage for European agriculture, so that it adjusts efficiently and effectively to the above-mentioned directional axes. It is estimated that the resources that will be made available for the support of Greek agriculture will be maintained around the 1997 level for the entire period of the Agenda 2000. Furthermore, additional resources from Structural Funds will be made available for the promotion of local development actions and initiatives. Consequently, the adaptation of Greek agriculture to the conditions of a global market, in a way that will secure competitiveness and sustainability, will depend on the efficient utilization of the available resources and especially on the emphasis that will be placed on the sustainable and integrated development of the Greek countryside (Damianos et al., 1994; OECD 1995).

## Redirecting Greek Agricultural Policy

Despite the existence of a CAP for all EU member states, there is always room for the implementation, in a supplementary but decisive way, of a national agricultural policy. This national policy need not be inconsistent with the common policy. It is well accepted and understood in Greece that the support of the agricultural sector constitutes a prerequisite and a priority for the overall economic and social development of the country. In other words, agricultural policy can and should be incorporated into the framework of a dynamic and comprehensive development strategy for the Greek countryside. This implies of course that Greek agriculture must undergo major adjustments in the face of challenges from an ever more competitive global environment. These adjustments are necessary because agricultural activity still plays a crucial role, especially in the regions of the country where agriculture constitutes the only source of income and employment.

The basic aims of a national policy could be: a) the creation of a vigorous and competitive agriculture, b) the viable and sustainable development of the countryside and, c) the ensuring of social cohesion and protection for the rural population. The major features of the sector that are impeding the accomplishment of these goals are many and well known: the rapid deterioration of the age structure of the rural workforce, the entire absence of agricultural education and training and of agricultural extension services, the uneven distribution of farmers' agricultural and total income, the overproduction of conventional products of mediocre quality, the imperfections in the marketing and distribution networks, the inadequate development of co-operative activities, the sluggish rate of improvement of rural infrastructure, the insufficient institutional support for agricultural and non agricultural economic activities of citizens in the countryside and, finally, the inability of the Greek state to adjust rapidly to changing socio-economic conditions in the countryside.

*Improvement of Competitiveness*

With regard to competitiveness, the national agricultural policy should concentrate primarily on young farmers and on the promotion and support of 'second generation' investments. This presupposes appropriate organizational and institutional support, which is not yet available. A major restructuring of state agencies that control and operate agricultural education, training and extension programmes is in process. The emphasis should be on the improvement of the effectiveness of the various training programmes on young farmers. The goal of substantially reducing production costs is not achieved with the adoption of measures like the subsidization of the means of production, which distort further an already distorted market, but rather with the promotion of measures and actions which ensure a real decrease in production costs. Such measures are the improvement of structures of marketing and processing of agricultural products, the reduction of the cost of agricultural land, the increased productivity of the workforce, the operation of a rational mechanism for the management of quotas, the general improvement of rural infrastructure (road networks, etc.) and, last but not least, the restructuring of the agricultural co-operative movement.

In the framework of a national agricultural policy, the problem of low competitiveness is handled mainly as a problem of commercial utilization of production rather than as a problem of non-competitive costs of primary production. In order to achieve this increased competitiveness, emphasis is given to investments that improve the quality of production, the timely and systematic adoption of new technologies and production techniques, the upgrading of agricultural training, education and extensions, the improvement of marketing strategies and the production and adoption of marketing innovations. A crucial factor for the promotion of products of high commercial value in national and international distribution networks is the establishment of stronger ties and connections between the producer of agricultural products and the processor and manufacturer of foodstuffs. In this context, increased competitiveness presupposes the establishment of an agency for the planning, certification and promotion of agricultural products of high added value and special quality specifications.

*Rural Development*

With regard to rural development, the aim is to maintain a quantitatively sufficient, economically and demographically viable, population in the Greek countryside, particularly where this is more difficult like in the mountainous areas, islands, less developed and border regions. In order to achieve this goal, programmes for integrated and multi-dimensional development must be implemented in every region, every prefecture and every local rural community. With these programmes, the development of other productive and social activities besides agriculture can be facilitated. In effect, one should always keep in mind the importance of multiple-job holding for rural development.

Targeted policies, differentiated by region, could lead to substantially lower inequalities in the Greek countryside. In terms of a policy for sustainable development, obstacles that deter farmers and particularly young farmers from remaining in rural areas must be lifted widely as soon as possible. Such obstacles are, for instance, the legal obstacles which prevent the full utilization of available agricultural land, the insufficient quantity and quality of technical support provided by the state, the lack of an appropriate institutional framework, the inadequate, and some times unacceptable, social and cultural conditions existing in the countryside, especially in areas that are remote from urban centres.

EU policy foresees measures and actions for agricultural development at a regional level and significant resources were made available within the confines of the Second Community Support Framework (Second CSF). Most of the resources were directed toward the improvement of infrastructure with small and large-scale works. So far, the Greek experience from the application of the Second CSF has shown that this plan suffers from complexity and fragmentation of the decision making process in both its legislative framework and implementation. Furthermore, decision-making is not transparent and in several occasions actions are in conflict with each other and quite often diverge from regional and/or local needs.

The target of rural development is not served adequately by the organizational, institutional and social realities of Greece, not given the proper importance for the improvement of social welfare and social benefits. The policy of local development should serve the declared objectives of agricultural development, such as the production of competitive agricultural products; at the same time it should serve the goal of social and economic cohesion as well as that of environmental and ecological stability.

The economic development of mountainous, island and less favoured areas of Greece, which cover almost 80 per cent of the land, attracts much of the attention in the discussions concerning the economic development of the country. The inability of these areas to sustain activities related to agricultural production can be attributed to the natural disadvantages they are subjected to and which, at the end, characterize them. In these areas, however, the continuation of agricultural activity and the maintenance of at least a minimum number of residents are necessary conditions for the preservation of their natural environment and the deterrence of further desertification.

Development policy for these areas is currently based on income supplements, increased investment incentives, (within the framework of the various social and structural programmes and regulations), and interventions through the various integrated development programmes and initiatives. The effective materialization of this package of measures runs up against the weaknesses and inadequacies that exist at the level of administrative and technical support, especially in the country's more remote areas. The lack of specialized staff is reflected in the quality of Regional Development Programmes that are being drafted, approved and employed, as well as in the irrational manner in which scarce resources are utilized. Interventions in the Third CSF and other community initiatives are expected to reduce the problems facing less developed areas and speed up development, under the condition effective control is exercised during the 'setting of priorities' stage and an in depth evaluation is conducted of the manner in which effective intervention is to be pursued.

*Social Protection*

Protection, social cohesion and economic welfare constitute the central goal of the national strategy for the rural population. The state secures retirement benefits and health care services and compensates farmers for damages caused by natural disasters. In this framework, the Organization of Agricultural Insurance is upgraded into an autonomous, self sufficient and viable insurance organization. At the same time the system is leading to higher pensions for every farmer, providing special care to female farmers and/or to the wives of farmers and a major upgrading of the health care services provided to all rural residents is pursued.

The need to improve the competitiveness of Greek agriculture goes hand in hand with the need for social cohesion. Given the serious unemployment problem of Greece, the major objective of national policy is the creation of new employment positions and opportunities in the countryside. This objective can be achieved by diverting resources to the countryside and making sure that rural population is not overburdened with the costs of structural adjustments. The incorporation of agricultural development into the wider framework of rural development policies in Greece, (contrary to the intensely sector-wide approach applied at present), is expected to be more efficient in achieving higher levels of economic development and social stability.

The integrated approach, with dominant geographic criteria, to the question of rural development introduces a new philosophy in the way state officials are pursuing rural development. Consequently, some of the political power lying today exclusively behind the support for agricultural products should be diverted toward structural policy and non-agricultural interventions in the rural areas. In this direction, it should be made clear that agricultural activity is a necessary but not a sufficient condition for the development of rural Greece and the maintenance of a flourishing and active population in the countryside.

## Epilogue

Recently, the Ministry of Agriculture made public its plan for the development of agriculture in the Third CSF, with an implementation period of 2000 to 2006. This is indeed an especially large and significant programme, with a total budget of EU, national and private contributions close to three trillion drachmas. The size of this programme becomes more evident when we consider that the corresponding level of the Second CSF, carried out in the period (1994 to 1999) was about one trillion drachmas. In contrast to the Second CSF, which was characterized by an intensely sector-specific approach, the new plan of the Ministry of Agriculture is upgrading significantly the integrated character of the support. In other words, the weight of the interventions seems to be moving toward what is commonly understood as integrated development of the countryside. This is definitely a move in the right direction.

It is worth mentioning that about 90 per cent of the funds in the agricultural part of the Second CSF were spent on the following four measures: a) income compensations for farmers in mountainous and less favoured areas, b) support for the improvement of the farm holding and the housing conditions of young farmers, c) financing of old and new irrigation works, and d) support for the improvement of processing and marketing of agricultural and forestry products. The main characteristics of the Third CSF are its wider selectivity and clear geographic dimension that favour intervention at a local and regional level, much more than the Second CSF did. It includes activities targeted at: a) the improvement in the age composition of the rural population, b) the education, training and extension services for farmers and residents of the countryside, c) the development and protection of natural resources and the environment, d) the development of agro-tourism and handicraft production, and e) the implementation of integrated programmes for local development.

Of course, in order for these measures and activities to bear fruit, not only is the better organization and staffing of the services of the Ministry of Agriculture necessary, but institutions and mechanisms must be created for the substantive participation of the rural residents in the implementation of the truly innovative actions being proposed. In other words, what is necessary is the creation of the proper structure of incentives that will stimulate the economic and social potential of the countryside and will make that the lever of self-motivated sustainable rural development.

## Note

1   This paper relies heavily on the Report by Damianos, D., Demoussis, M., Kasimis, C. and Skuras, D., 'Third Community Support Framework 2000-2006: Reorienting Greek Agricultural Policy', Department of Economics, University of Patras, December 1997, (in Greek).

## References

Damianos, D. et al. (1994), *Pluriactivity in the Agricultural Sector and Policies for Rural Development in Greece*, Foundation for Mediterranean Studies, Athens, (in Greek).

Damianos, D. et al. (1998), *Greek Agriculture in a Changing International Environment*, Ashgate, Aldershot.

Demoussis, M. and Sarris, A. (1988), 'Greek Experience Under the CAP', *European Review of Agricultural Economics*, Vol. 15(1), pp. 89-107.

Georgakopoulos, T. and Paschos, P. (1985), 'Greek Agriculture and the CAP', *European Review of Agricultural Economics*, Vol. 12(4), pp. 247-263.

Maraveyas, N. (1992a), *Agricultural Policy and Economic Development in Greece*, Nea Synora, Athens, (in Greek).

Maraveyas, N. (1992b), *The European Integration Process and Greek Agriculture in the 1990s*, Papazisis, Athens, (in Greek).

OECD (1995), *Better Policies for Rural Development*, OECD Documents, The Proceedings of the High Level Meeting of the Group of the Council on Rural Development.

Ostrom, E. (1996), 'Incentives, Rules of the Game, and Development', in M. Bruno and B. Pleskovic (eds), *Annual World Bank Conference on Development Economics 1995*, The World Bank: Washington, D.C.

Sapounas, G. and Miliakos, D. (1996), *Greek Agriculture in the Post-War Era: The Effects from Acceding the European Union and Lessons for the Future*, Studies of Agricultural Economics, No. 51, Agricultural Bank of Greece, Athens, (in Greek).

Sarris, A. et al. (1996), *The Uruguay Round Agreement for International Trade and its Impact on Greek Agriculture*, Foundation of Economic and Industrial Studies, Athens.

Tracy, M. (1997), *Agricultural Policy in the European Union and Other Market Economics*, 2nd edition, Agricultural Policy Studies, Genappe.

# Index